SUPPLY CHAIN RISK MANAGEMENT: COMPETING IN THE AGE OF DISRUPTION

CERM Academy Series on Enterprise Risk Management (ERM)

Greg Hutchins PE CERM
CERMAcademy.com
Gregh@CERMAcademy.com
503.233.1012
800.COMPETE

HOW TO ORDER:

Cost is $89.00 per copy plus $6.00 Shipping/Handling in U.S. Offshore orders are based on the form of delivery. Quantity discounts are available from the publisher.

Quality Plus Engineering 503.233.1012
4052 NE Couch   800.COMPETE
Portland, OR 97232   800.266.7383
USA   GregH@CERMAcademy.com

For bulk purchases, on company letterhead please include information concerning the intended use of the books and the number of books to purchase.
© 2018 rev 1 - Quality + Engineering/CERM Academy

TABLE OF CONTENTS

CHAPTER 0
PREFACE

Supply Chain Risk Management is the third edition of this book.

The title of the first purchasing book I wrote twenty years ago was: **Purchasing Strategies for Total Quality**. The book focused on purchasing strategies involving quality, delivery, and cost.

Supply Management Strategies for Improved Performance was the title of the second iteration of this book. Many things happened from the first to second edition. We expanded beyond quality. Purchasing matured to supply chain management. There was more emphasis on business performance and innovation.

The second title described new sourcing priorities. In the second book, we covered the state of supply management, presented best practices, and provided guidelines on how to jumpstart a supply chain initiative.

The title of the third iteration of this book is **Supply Chain Risk Management: Competing in the Age of Disruption**. In this revision, we focus on the following:

- Illustrate how supply chain management is migrating to Supply Chain Risk Management (SCRM).

- Demonstrate how SCRM objectives align with the organization's strategic business objectives.

- Describe how to move beyond a price focused relationship to a value-added relationship with suppliers based on risk.

- Integrate the disparate elements of SCRM into a competitive business system.

- Explain why Risk Based Problem Solving (RBPS) and Risk Based Decision Making (RBDM) are the future of SCRM. Examples are offered throughout the book.

- Describe how to select and develop suppliers based on risk criteria.

- Demonstrate how to use ISO 31000 risk management framework as the foundation and architecture for SCRM.

- Illustrate how supply chain risk-controls are architected, designed, deployed, and assured.

- Prepare supply chain, quality, engineering, and operational excellence professionals for their emerging risk roles, responsibilities, and authorities.

FROM PURCHASING TO SUPPLY CHAIN RISK MANAGEMENT

Knowledge is power.
Proverb

SCRM is comprised of tools and techniques, such as risk management, operations excellence, lean, just-in-time, Six Sigma, and Enterprise Risk Management (ERM). Each of these innovative philosophies has a different functional home. Quality owns Six Sigma. Materials management owns inventory control. Planning owns demand forecasting. Manufacturing owns production control and lean production. Engineering owns lean design. Purchasing owns supply relationships. Compliance owns ERM.

Is there one central coordinating and communicating function for supply risk issues? No. Does there need to be a group that owns SCRM if organizations are going to coordinate and resolve supply chain risk issues?

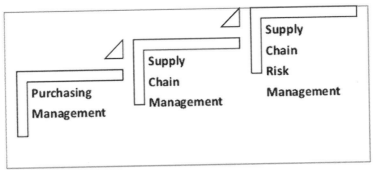

*Figure 1: **From Purchasing to SCRM***

Yes. Where should it reside? More organizations are developing operational risk groups.

TRADITIONAL PURCHASING

Purchasing started as an order taking and back office department, largely transactional and clerical. This worked for many years. Especially in large organizations, outsourcing and offshoring have forced the purchasing department to mature from a tactical to a strategic point of view.

However, many organizations still have purchasing departments with buyers. Some of these departments have matured to supply management. Others have not. And, this is a conundrum for many organizations, should they add the additional responsibilities on the current function, create a separate supply chain organization, integrate many functions into one, or even outsource the function. There is no simple answer. However, this is a critical discussion because the next iteration for the function we believe will be Supply Chain Risk Management. The evolution from Purchasing to Supply Chain Risk Management is illustrated in the above figure.

SUPPLY CHAIN RISK MANAGEMENT

In terms of total dollar amount, external suppliers provide a significant portion of a manufacturer's product. For U.S. firms, 80% or more of the final price of a product is the cost of purchased goods. In Japan, it can even be higher. For these reasons, SCRM is critical to a company's competitiveness and success.

The idea of SCRM has been around for about 10 years. It is only in the last 5 years, that it has assumed more importance. The good news, we are still in the early stage of SCRM adoption and deployment.

ELEVATION OF SUPPLY MANAGEMENT
Smaller organizations tend to look at supply management as a purchasing function that is cost driven. The purchasing function until a dozen years ago was a director level function.

As more end-product manufacturers developed an outsourcing business model, supply management has been elevated to a senior vice president or even higher position within many organizations.

In terms of how the supply chain group works, Supply Chain Risk Management teams are cross-functional. Cross-functional team representatives may come from quality, production, purchasing, engineering, planning, and other interested stakeholders.

CONVENTIONAL USAGE IN THIS BOOK
A supply chain can be massive involving many thousands of suppliers throughout the world. So when people refer to a supply chain, it is critical to scope what is meant by the concept. Are we talking about first-tier suppliers including their suppliers (second-tier and lower)? If we are, the scale of the supply chain becomes huge and unmanageable. For the sake of clarity, we will generally refer to first-tier or key supply-partners.

To avoid confusion, we will refer to the brand owner as the end-product manufacturer and first-tier and downstream suppliers by their location in the value stream or supply chain.

In this book, we use the term final-customer to reflect the end user or the final purchaser of the product or service. The customer of a supplier is called the end-product manufacturer, the company whose logo or brand is on the package. The end-product manufacturer is sometimes called the Original Equipment Manufacturer (OEM). A first-tier supplier may also be

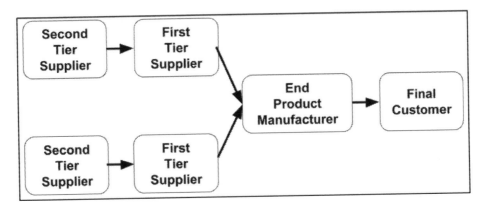

Figure 2: **Customer - End-Product Manufacturer - Suppliers**

the customer of the second-tier supplier and so on down the supply chain. These concepts are illustrated in the above figure.

When we use the expression 'customer-supply', this refers to the relationship between the end-product manufacturer (brand owner) and its suppliers. Or expressed another way, the customer in the 'customer-supply' expression is the end-product manufacturer relating with and interacting with its first-tier supplier (s).

Why do we not use the term Original Equipment Manufacturer (OEM)? There is confusion, ambiguity, and lack of consistent use on how the term is used. An OEM can be the end-product manufacturer, first-tier supplier, brand owner, branded component/subassembly/assembly, rebranded product supplier, or even aftermarket supplier.

WAYS TO GET STARTED

Some readers will purchase this book and wonder how to get started with a SCRM initiative. Others will purchase this book with supply risk management programs already in place and want to know what they can do to move to the next level. As well, others will purchase this book with other requirements.

In this book, we offer several options for starting a SCRM program, specifically:

- Using ISO 31000 or similar risk management framework to start, design, deploy, and assure a SCRM program. Please visit page 121 to view guidance.

- Using a project based approach to start or mature a SCRM program. Please visit page 227 to view guidance.

- Using a supply certification approach to start or mature a SCRM program. Please visit page 257 to view guidance.

- Using operational excellence tools approach to start or mature a SCRM program. Please visit page 351 to view guidance.

Each supply chain is different. Thus, SCRM has to be architected, designed, deployed, and assured based on context, products, and services being delivered and procured; customer, stakeholder, and interested party requirements; and risk maturity and capability of the supply chain key players.

RBT, RBPS, AND RBDM EXAMPLES
As well, the critical elements of SCRM: Risk Based Thinking (RBT), Risk Based Problem Solving (RBPS), and Risk Based Decision Making (RBDM) must be designed and tailored to the supply chain. RBT, RBPS, and RBDM are fully detailed, integrated, and illustrated in this SCRM book.

This book provides general guidelines and best practices. We suggest first reading or skimming the book from beginning to end. Note the specific pages or chapters listed above on how you want to start the SCRM program then pick and choose the practices that would work for your organization and supply chain.

SUPPLY CHAN RISK MANAGEMENT READERS
The readers of this book can come from different functions:

- Executive management.

- Operations.

- Supply management.

- Purchasing.

- Quality.

- Engineering.

- Production.

- Manufacturing.

- Process sectors.

- Healthcare.

- Service.

- Financial.

- Cyber Security.

- Information Technology.

- Government.

The titles of the readers benefiting from this book may include:

- Chief Supply Management Officers.

- Vice presidents.

- Directors.

- Managers.

- Supervisors.

- Consultants.

- Lean professionals.

- Operations managers.

- Design engineers.

- Product development engineers.

- Quality engineers.

- Operations managers.

- Purchasing professionals.

- Supply managers.

- Production planners.

- Manufacturing and production managers.

- Marketing professionals.

- Distribution professionals.

- Inventory managers.

- Logistics professionals.

- Students.

SUMMARY

Risk management is the biggest issue for supply management teams right now and feeds directly into the success of supply chains.[1] As well, supply management is moving from a transactional and product based business model to an integrated Supply Chain Risk Management model, which is abbreviated as SCRM throughout this book.

This book provides general guidelines for architecting, designing, deploying, and assuring the adequacy and effectiveness of supply chain risk-controls. The book provides general scenarios, approaches, processes, best practices, and tools that have worked well with our clients and other organizations. This book does not provide prescriptions to developing a specific approach to SCRM.

We believe in continuous innovation. If you have comments or thoughts on how to improve this book, please call us at 800.COMPETE

(800.266.7383). We want this to be the best supply chain risk product around. Thank you for purchasing this book.

Greg Hutchins PE CERM
Gregh@CERMAcademy.com
503.233.1012 or 800.COMPETE
CERMAcademy.com
800Compete.com
WorkingIt.com

CHAPTER 1:
MANAGING IN VUCA TIME

WHAT IS THE KEY IDEA IN THIS CHAPTER?

Recent reports indicate that from 2014 to 2015, there was a 118% increase in disruptive supply chain events.[2] Why? We live in a world and time when:

> "Global political risk is at the highest level since the height of the most severe crises in the Cold War (the Cuban crisis, Yom Kippur war, etc.)."[3]

What does this mean? Uncertainty is the new normal. *Forbes Magazine* and KPMG in their recent report stated:

> "Scan the business news and it can seem like the way almost everything works is being rewritten: how we interact, how we buy and sell, how we make things, how we get from place to place. ... We are amidst the largest era of institutional change in the history of our planet."[4]

VUCA IS TODAY'S SCRM PARADIGM BUSTER

There is no finish line.
Nike Corporation motto

Just think of all the changes occurring in your organization. New regulations. Changing customers with new product preferences. New competitive product offerings. New technologies. New branding. New business platforms. New business models. Compressed lifecycles. Change is no longer gradual. Change is rapid and abrupt.

This chapter is about the disruptive changes occurring in supply chains, management, and even all professions (quality, engineering, supply management, etc.). As well, we believe general management is evolving into risk management.

In the previous edition of this book, we believed competitiveness was the supply chain paradigm buster as distilled in the below quotation:

> "Capitalism, competitiveness, dot.com profitability, and innovation are today's paradigm busters. They are what drive today's New Economy. Capitalism is triumphant in almost every activity in every corner of the world. National boundaries are disappearing as goods and services move freely. Competitiveness and innovation are continually destroying the current business paradigm. Just as a business feels it has found its niche and understands what is going on, supply management and sourcing rules change again."[5]

WHAT IS VUCA?

Supply Chain Risk Management reflects the major changes occurring in today's economy and marketplace. What do I mean? ISO 9001 is an international Quality Management System standard. When ISO 9001 came out in 1987, the focus was on consistent quality. It was a great idea for its time. Quality management, quality assurance, and quality control were essential to a company's competitiveness.

More than 30 years later, the emphasis is how to compete and innovate in a time of Volatility, Uncertainty, Complexity, and Ambiguity (VUCA) or what we call VUCA time. Examples of VUCA are all around us. As an example, a Black Swan event is a big bang or VUCA disruptor to an organization. A Black Swan is a high consequence and low likelihood event that can disrupt a company's supply chain. A Black Swan can be climate change, hurricane, loss of a final-customer, or loss of a critical supplier. As you can imagine, you are going to hear a lot about VUCA in this book.

In a world with interconnected supply chains, business and operational risks surface often. Why? Unexpected events occur more frequently. For example, earthquakes, tsunamis, oil spills, increasing global warming, and

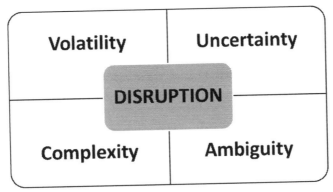

Figure 3: 4 Elements of Disruption

loss of a single-source of products or services can rapidly impact a company and its supply chain potentially undermining the company's reputation and its financials.

Exposure to global risks is increasing and the consequences are ever more threatening. Also, something as simple as 100-year events are occurring more frequently. VUCA is the driver of all disruption as shown in the above figure. That is why we say: "we live in the age of VUCA."

VUCA can impact downside risk as well as upside risk or opportunity risk. Opportunity risk may include: mergers, acquisitions, product development, innovation, investment opportunities, infrastructure build-out, and the costs of borrowing for capital projects. Opportunity risk is becoming so critical that investors are demanding higher levels of risk-assurance and transparency from bond rating agencies such as Standard & Poor's, which now uses Enterprise Risk Management to evaluate companies. Newspapers also use the expression 'the risk of an overheating economy', which is upside risk.

VUCA EXAMPLES
Below are several examples of SCRM VUCA:

- **Time VUCA.** Time compression is impacting supply chains in many sectors. In retail, fashion design used to involve four seasons. Now, fashion design firms provide new dresses weekly to customers.

- **Geographic VUCA.** From supply chain perspective, disruption is often a country specific issue. For example, in India the drivers of disruption are regulatory, money issues, dependence on a single-source, language issues, security issues, water availability, and even technology differences.[6]

- **Climate VUCA.** Supply chain disruptions can also come about from unexpected sources. Climate change is causing droughts in India, Australia, United States, China, and many parts of the world. In India, the drought is impacting 300 million people or a quarter of India's population after 2 monsoons delivered low rainfall. Water supply disruptions can cripple domestic production and services for global supply chains requiring water and power.[7]

VUCA DRIVERS

If anything is certain, it is that change is certain. The world we are planning for today will not exist in this form tomorrow.
Phil Crosby

How does an end-product manufacturer become competitive in VUCA time? This is the key question in this chapter and throughout this book. We can offer suggestions. However, each organization is unique so there is no simple answer.

VUCA DRIVERS

Something or someone has to be the impetus and catalyst for SCRM change like a new competitor, increasing/changing customer expectations, changing end-product manufacturer requirements, higher regulatory oversight, product liability litigation, or a consequential supply event.

The following are early warning signs or precursors that supply chain disruption at either the end-product manufacturer or within the supply chain is coming:

- Changing competitive environment.

- Development of new business models by competitors.

- Black Swan or unexpected events with catastrophic impacts.

- Public safety threats or hazards.

- Pending litigation.

- Mounting operational losses.

- Regulatory action.

- Negative final-customer surveys.

- Product recalls.

- Low internal employee satisfaction.

- Low final-customer satisfaction.

- Slow or very fast company growth.

- Mergers, acquisitions, reengineering, or other transformations.

- Inflexible or fat supply chain.

- No executive management commitment for SCRM.

- Strong departmental silos.

- Low profit margins.

- Poor intra and inter customer-supply communications.

- Supply product containment issues.

- Unclear supply chain goals and metrics/objectives.

- No emphasis on SCRM training or learning.

- Resistance to change.

FAVORING THE STATUS QUO

As the velocity, the rate of change, of technology increases, organizations must adapt and anticipate these changes. SCRM is becoming the favored approach to respond and even anticipate VUCA and other disruptive changes.

VUCA challenges the basic premise and purpose of the end-product manufacturer and its supply base. Companies, organizations, and institutions
have a body of culture, values, principles, and infrastructure that seem to
value the status quo in direct challenge to the acceleration of technology,
disruption, and the need for competitiveness.

Supply chain disruptions can also create a domino risk effect. For example
in a lean production environment, there is little buffer inventory and little
time for manufacturing to recover from a disruption to the supply chain resulting in missed end-product manufacturer deadlines. With operational
excellence best practices focusing on Lean Six Sigma, operating managers
are often not prepared and do not have the tools to effectively manage
supply chain risks.

SUPPLY CHAIN DISRUPTION IN THE NEWS
Samsung phone batteries. Chipotle food. Even Ben & Jerry ingredients
have had recent quality problems that involve billions of lost dollars resulting in reputation loss and ultimately brand dilution.

As is seen with Takata airbags and Samsung phone batteries, technology
is a major VUCA supply chain disruptor. Robots, 3D printers, and instant
communication allow products to be made just-in-time in units of 1 to satisfy
final-customer specific requirements. 3D printing allows small companies
to produce precise dimensional products with specific tolerances based
upon requirements. Robots in the middle of a store can pack and ship
products. Instant communications allow electronic invoicing and payment.
Radio Frequency Identification (RFID) tags change labeling, data collection, and warehouse management so unique products can be identified and
traced from production to the final-customer.

Let us look at some recent examples of disruptors in the news:

CLIMATE CHANGE RISK
Climate change is increasing supply chain risk. Climate change results in
CO_2 emissions. Micro climates form where rainfall may increase in areas
with abundant water and decrease in areas facing aridity. Global coastal

cities become prone to flooding. Food chains are disrupted. Water supply disruptions are more common as a result of climate change.

Over 80% of the world's manufacturing occurs in coastal areas, that are impacted by coastal flooding, earthquakes, hurricanes, typhoons, and tsunamis. For example, more than 300 million people in India or a quarter of its population were impacted because two successive monsoons failed to deliver sufficient water.

CYBER SECURITY RISK

Cyber security risk is among the most critical risks to the supply chain:

> "Cyber and data privacy breaches are perceived to be the largest threat to the stability of transport and logistics, with the sector facing potential breaches of $2 trillion by 2019."[8]

The reality of a cyber-crime is not if, but when, and how damaging it will be. The inevitability of a cyber-attack has forced end-product manufacturers to reassess every element of the end-product manufacturer from product design, outsourcing, and servicing the product.

Cyber security risks will increase exponentially over the next five years. The challenge is that most products have some type of software. So more often, suppliers are providing critical products with embedded software that has not been sufficiently quality controlled. Huge problem. Counterfeit components, malicious software, or hacktivists can disrupt the supply chain or at a minimum degrade functionality. Just look at susceptible technology in today's automobile: automatic collision avoidance, lane-keeping assistance, adaptive cruise control, distance maintenance, front-car collision-avoidance systems, and 360^0 cameras.

HACKER ATTACK RISK

Hackers can be hacktivists or nation states. Hackers more often are exploiting the digital supply chain to add malware or malicious code that hides in the software company's core software, installation, or patches. As a result of the Internet of Things (IoT), more hard products have built-in software that are Internet accessible. Supply chain, cyber-attacks focus on these soft targets such as smart refrigerators that are breachable with relatively little effort. As well, the benefits of hacking compared to the amount of effort are high. A hacker can hijack final-customer personal information, core Intellectual Property, financial information, and other core assets relatively easily.

High profile hacker attacks also have become common, just look at Equifax, Securities and Exchange Commission, Deloitte, and many others. This is the new normal for governments and end-product manufacturers. Most importantly, these attacks have unimagined consequences. Aside from losing invaluable information due to the breach, now executive management heads are beginning to roll. The Chief Information Security Officer and even the Chief Executive Officer may be fired.

Breaches can to have organizational consequences. The perception is that if senior executives cannot control the organization's most valuable assets, such as Intellectual Property, then does the business know what it is doing. Investors pay attention to their investments and want to know that they are secure. That is why internal and third-party cyber security have become paramount issues with end-product manufacturers and government.

PACEMAKER RISK

A few years ago, a TV program showed a killer hacking the wireless monitoring capability of a pace maker and giving it commands to accelerate electrical impulses to the heart and killing the patient. Great idea for a fictional story. However, these vulnerability and threat risks are now real.

Can you imagine wearing a heart monitor and then find the monitor signaling a very high and dangerous heart rate? And, what if the monitor could send an electrical impulse to your heart based on its monitoring? But, you did not need the extra electrical impulse. Your heart was OK. But, someone had hijacked your medical device and could now control it. Sound far-fetched? Not really. This is the brave new world of the Internet of Things (IoT).

Abbott announced that it was voluntarily recalling 465,000 pacemakers due to cybersecurity vulnerabilities in the device. In other words, the TV program was prescient. The program anticipated what could go wrong with a medical device, specifically allowing the attacker to modify the devices' pacing commands or cause the battery to prematurely deplete.[9]

INTERNET OF THINGS RISK

More products have electronics and smarts built into them. Something as simple as a Light Emitting Diode (LED) light has a microcircuit in its base, which is controlled by software. The LED light is part of the smart house where heating, lighting, and door access are controlled via a smart phone over the web. So, the LED light's electronic controls connect with the web, which can tell a burglar or another bad person that no one is at home and how to open doors. These smart devices are now hackable, accessible, and exploitable by people with bad intentions.[10]

The Internet of Things (IoT) involves new technology where toasters and even washing machines have an Internet Protocol (IP) address so machines can talk to each other. The IoT is expected to grow exponentially. This will have a profound impact on supply chains. Products will be able to share information, become self-aware, communicate, and even correct themselves. At the system level, global supply chains will be able to detect, and remediate supply chain risks.

However as with most technology, there is a double-edged sword. The IoT will create entirely new categories of cyber security threats and challenges. These products will have an IP address that will potentially be hackable. A critical question is whether competitors, nation states, or bad people will use these products to penetrate supply chains to steal critical Intellectual Property and use the information for illegal purposes. We are already seeing more supply chain security breaches and malicious threats.

3D PRINTING RISKS

Low-cost suppliers were the original driver to outsourcing and offshoring. However, 3D printing at a local production site removes the benefit of cheap offshore labor and high transportation costs.

3D printing is the process and technology to provide unique products based on special requirements. 3D printing is a technology that will reshape design, manufacturing, and supply management. Products with complex shapes are made on-demand based on unique end-product manufacturer requirements. Production times are reduced. New and exotic materials are used to produce new products.

3D printing also allows for simpler products, fewer parts, and simpler design/production processes. Tiers of suppliers are eliminated. Printing is done next to the assembly and test plant. Printed products will be produced on-demand. Inventory will not exist or be minimal.[11] It is also anticipated that 3D printing will move to the atomic level in medicine as labs will provide 'tissues on-demand' or 'orthopedics on-demand.'[12]

BMW RISKS

BMW the German carmaker was not able to complete the production of a new luxury vehicle because the car maker did not have a small but critical component used to make electrical assistive steering systems. Plants in South Africa and China were impacted.

The auto industry's reliance on lean and just-in-time systems means that a break or disruption can travel up the supply chain very quickly causing substantial losses for the end-product manufacturer. BMW now is working with key suppliers to develop in-plant solutions for managing supply chain risk.[13]

HYUNDAI RISKS

What happens when there is an even minor disruption in a lean supply chain. Almost weekly, we are hearing that the entire end-product manufacturer may have to stop production.

Hyundai is South Korea's biggest car maker. Recently Hyundai Motor had to shut operations of its four Chinese plants because a local supplier refused to provide products because of delayed payments. The supplier of plastic fuel tanks stopped 4 assembly plants that produced up to 1.35 million vehicles in China.[14]

SAMSUNG BATTERY RISKS

Samsung makes world-class smart phones selling at a premium price. Samsung smart phones generate substantial revenue but also add luster to the Samsung image, which transfers to brand equity. This was the case until Samsung Galaxy Note 7 fires caused by its lithium ion battery resulted in a global recall. The reports of fires were so dire that airline regulators throughout the world issued rules against their use on airplanes.

What happened? Samsung allocated 60% of its battery orders to a single-supplier. Samsung did not sufficiently test the battery. Samsung did not have a product recall plan. Samsung did not have sufficient alternate suppliers to fulfill demand posed by the recall and additional orders. Samsung's recall of its Galaxy Note Seven smart phone is a textbook example of poor SCRM.[15]

BEN & JERRY'S RISK

End-product manufacturers are often not prepared for VUCA disruption. Look at Ben & Jerry's, which is a premium ice cream producer. The company is an environmentally conscious, end-product manufacturer selling a premium product based on its reputation.

One of its key selling differentiators was selling kid-friendly, nutritious, and non-GMO (genetically modified organism) ice cream. Well, it was discovered that its ice cream had traces of glyphosate in its products, which is an herbicide and weed killer ingredient. Ben & Jerry's was not aware how or when the herbicide entered its products. However, the company's reputation is now at risk especially when the public explanation from the company was "the poison in question is in such small quantities that it is not going to hurt anybody" explanation.[16]

RETAIL RISK

The retail channel is disrupted in almost all segments because of what is often called the Amazon effect resulting in retail apocalypse. In many countries, malls, anchor retail, and small business retail shops are empty. Why? More if not most retail buying is online through Amazon or Alibaba in China. Retail customers can now buy directly from a supplier thus eliminating the middle company and distribution costs.

Ecommerce will soon be the majority of retail sales. According to experts:

> "The store of the future will become the most powerful media channel available to a brand, offering customer experiences that are the

most profitable product a retailer can sell. But to get there, retail as
we know it must die."[17]

Retail shops and anchor department shops will have to reinvent themselves and innovate. This change management will be difficult to most end-
product manufacturers because they will have to reinvent their culture, redesign their business models, and accommodate to the new purchasing
normal:

> "Everyone is talking about the need for disruption, innovation and
> change, yet most (retailers) will stop well short of actually doing an-
> ything about it. Many retail brands talk about game-changing inno-
> vation but what we see are lukewarm iterations of existing concepts
> and old ideas. Retailers, it seems, lack the will or sense of urgency
> to effect significant and radical change. [18]

TRADE WAR AND RISK CONSEQUENCES

More countries want domestic content products to fulfill promises of local
jobs. The U.S. and many exporting countries including China, Mexico,
Canada and most developed countries are imposing tariffs on imported
products. Risks can arise with known and unknown consequences as a
result of these requirements:

- **Direct Consequences:** "The raw material import costs for U.S. pro-
ducers of machinery and cars may rise. Therefore, foreign-manu-
factured may become cheaper for U.S. consumers. Consequently,
manufacturing companies in the U.S. may see higher costs of
goods sold and reductions in non-domestic sales."[19]

- **Indirect Consequences:** "Production and price changes will require sourcing changes. Sudden shifts that affect supply chains
may impact quality and availability, since some companies may en-
counter issues when scrambling to reduce production in some
places and ramp it up in others. Operational risk assessments will
help identify areas where change can have a positive or negative
impact." [20]

SUMMARY
The world has become an international marketplace. In this marketplace, products from Europe, the Pacific Rim, and South and North America compete for the final-customer's buying dollar. The money spent on a product by a customer means the end-product manufacturer that produced the product or the supply chain that delivered the product will have the capital to maintain and expand its operations. It also means the company will keep its people employed making more products to sell in the international marketplace.

VUCA factors, some would say opportunities and risks, are increasing the importance of SCRM. These factors include political risk, global economy, customer demand changes, end-product manufacturer requirement changes, low-price competition, environmentalism, consumerism, technological changes, and product lifecycle reduction.

Occurring simultaneously, customers are demanding high-quality products and services that are delivered cheaply, quickly, and courteously. Products and services are designed to satisfy national and local requirements. Time-to-market has become a critical success factor.

In a recent article, the author maintains that the entire premise of global and domestic sourcing may have to be reassessed due to technology and other disruptors. The author makes a good point since much of traditional purchasing, supply management, quality engineering, and customer-supply auditing is still reactive – after an event has occurred.

NEXT CHAPTER
In the next chapter, we discuss in more detail Supply Chain Risk Management in VUCA time.

CHAPTER 2:
SUPPLY CHAIN RISK
MANAGEMENT (SCRM)

WHAT IS THE KEY IDEA IN THIS CHAPTER?

Companies have experienced supply chain disruption is surprisingly high. An Aberdeen survey several years ago revealed that 99% of companies had a supply chain disruption over the previous year.[21]

Supply chain disruptions can create a domino effect on the availability of products and services. In a lean production environment, there is little buffer inventory and little time to recover if manufactured products have nonconformances resulting in missed end-product manufacturer deadlines leading to low final-customer satisfaction.

KEY SCRM DEFINITIONS AND CONCEPTS

All business proceeds on beliefs or judgments of probabilities, and not on certainties.
Unknown

Developing a common understanding of key SCRM definitions and concepts is the first challenge for many end-product manufacturers.

Unfortunately, there are many definitions of risk. Risk is the occurrence of an unwanted event. It is a deviation or variance from a target or norm, which can be a technical specification, contract, or engineering drawing.

WHAT IS RISK?

So, what is risk? Several definitions of risk include:

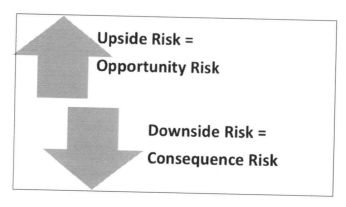

Figure 4: **Upside & Downside Risk**

An academic definition is:

> "Anything that hinders the flow of material, money and information can be identified as risk. Risk occurs due to uncertainty of future events. Risk may be internal or external. Risk may be short-term or long-term."[22]

The Institute of Internal Auditors (IIA) defines risk as:

> "... the probability that an event or action may adversely affect the organization or activity under review." [23]

International Organization of Standardization (ISO) 31000 standard on risk management defines risk as:

> "effect of uncertainty on objectives."[24]

In this book, we offer another definition of SCRM risk as illustrated in the above figure. We define risk as:

> "Value creation is upside or opportunity risk. Value detraction is downside or consequence risk. Anything that detracts from real and/or perceived value is a downside risk to a customer, stakeholder, or interested party. The goal of any organization is to optimize value creation and to minimize or control value detraction."

WHAT IS RISK MANAGEMENT?

Managers do not want uncertainty, which is a critical element of VUCA. Uncertainty results in risk. Risk management is evolving into a critical skill of all managers. Here are a few definitions of risk management:

ISO 31000 defines risk management as:

> "… coordinated activities to direct and control an organization with regard to risk"[25]

Business Dictionary defines risk management as:

> "The identification, analysis, assessment, control, and avoidance, minimization, or elimination of unacceptable risks. An organization may use risk assumption, risk avoidance, risk retention, risk transfer, or any other strategy (or combination of strategies) in proper management of future events." [26]

U.S. Department of Defense defines risk management as:

> "the risk that an adversary may sabotage, maliciously introduce unwanted function, or otherwise subvert the design, integrity, manufacturing, production, distribution, installation, operation, or maintenance of a system so as to surveil, deny, disrupt, or otherwise degrade the function, use, or operation of the system."[27]

WHAT IS A SUPPLY CHAIN?

Now, let us look at common definitions of a supply chain. You would be surprised to read that there is no universally accepted definition of a 'supply chain.'

APICS, the Educational Society for Resource Management, has one of the better definitions and defines 'supply chain' as:

> "a system of organizations, people, technologies, activities, information and resources involved in moving materials, products and services all the way through the manufacturing process, from the original supplier of materials supplier to the end customer."[28]

```
CONTEXT: Supply Chain Risk Concerns (Deloitte Survey)

•  Executives are concerned about risks to the extended value chain in-
   cluding outside suppliers, distributors, end-product manufacturers, and
   customers.

•  Margin erosion and sudden demand changes cause the greatest im-
   pacts to the heart of the business.

•  End-product manufacturers face a wide variety of supply chain risk
   challenges including collaboration, end-to-end visibility, and justifying
   investment in supply chain risk programs.

•  End-product manufacturers lack the latest SCRM tools.

•  SCRM is not always considered effective.[30]
```

Another 'supply chain' definition is:

> "A group of companies connected loosely, all collaborating on the
> same goal: efficient and economical product delivery. Or, the set of
> order entry and order fulfillment related physical interactions con-
> necting a company and its customers and suppliers."[29]

WHAT IS SUPPLY CHAIN MANAGEMENT?

Supply Chain Management is another concept that can create confusion:

APICS defines Supply Chain Management as:

> "the design, planning, execution, control, and monitoring of supply
> chain activities with the objective of creating net value, building a
> competitive infrastructure, leveraging worldwide logistics, synchro-
> nizing supply with demand, and measuring performance globally."[31]

International Tax Review defines supply chain management as:

"Supply chain management considers the management of materials and products from suppliers through all internal operations, including distribution to the customer. It is designed to:

- Optimize network and material flow.

- Reduce costs and cash consumption.

- Increase speed.

- Streamline, align, and focus information flow.

- Streamline and refocus the organization from functional and national to process and cross-border."[32]

WHAT IS SUPPLY CHAIN RISK MANAGEMENT

Several academic definitions include:

"The implementation of strategies to manage both every day and exceptional risks along the supply chain based on continuous risk assessment with the objective of reducing vulnerability and ensuring continuity."[33]

"Supply risk is defined as the probability of an incident associated with inbound supply from individual supplier failures or the supply market occurring, in which its outcomes result in the inability of the purchasing firm to meet customer demand or cause threats to customer life and safety."[34]

The U.S. Federal government defines SCRM as:

"Management of risk to the integrity, trustworthiness, and authenticity of products and services within the supply chain."[35]

"Supply Chain Risk Management is the process of identifying, assessing, and neutralizing risks associated with global and distributed nature product and service supply chains."[36]

Supply Chain Risk Leadership Council (SCRLC) defines supply chain risk as:

> "as the likelihood and consequence of events at any point in the end-to-end supply chain, from sources of raw materials to end use of customers" ... "as the coordination of activities to direct and control an enterprise's end-to-end supply chain with regard to supply chain risks."[37]

Mitre Corporation defines supply risk management as:

> "... the discipline that addresses the threats and vulnerabilities of commercially acquired information and communications technologies within and used by government information in weapons systems. Through SCRM, systems engineers can minimize the risk to systems of components obtained from sources that are not trusted or identifiable as well as the inferior materials or parts."[38]

KNOWN AND UNKNOWN RISKS

Another way of viewing supply chain risk is through the following risk lens:

- **Interdependent risks.** These risks are connected with other risks through business relationships, software, or other SCRM links.

- **Cascading risks.** These risks ripple across, up, or down a supply chain. For example in a lean supply chain, the loss of a fourth-tier critical part supplier can impact the third, second, and first-tier suppliers since they may not have the requisite part.

- **Avoidable risks**. Avoidable risks are those a company can prepare for. When a hurricane was barreling down to Florida, the weather forecasts provided at least a week's warning to big box retailers such as Home Depot. These companies stocked up their distribution centers, warehouses, and retail stores with the right products for people to sustain the disaster. This meant pre-staging supplies from suppliers to warehouses to retail stores in areas of the highest risk of disruption.[39]

- **Unavoidable risks.** Unavoidable risks are those that a company can not anticipate or prepare for. These Black Swans may be an earthquake or even a tsunami.

SUPPLY RISK MANAGEMENT PARADIGM SHIFTS

Every breakthrough business idea begins with solving a common problem. The bigger the problem, the bigger the opportunity. I discovered a big one when I took apart an IBM PC. I made two interesting discoveries: The components were all manufactured by other companies, and the system that retailed for $3,000 cost about $600 in parts.
Michael Dell

Paradigm (pronounced pair-a-dime) has become a cliché. Joel Barker in **Paradigms: The Business of Discovering the Future** defines a paradigm as:

> "...a set of rules and regulations (written or unwritten) that does two things: 1. It establishes or defines boundaries and 2. It tells you how to behave inside the boundaries in order to be successful."[40]

The word 'paradigm' comes from the Greek and means a pattern or model. A paradigm is the way people perceive their world. It can mean a world of difference. Fish perceive their world through water. We perceive our world through air.

Often, paradigms are defined in terms of a game. A game has a set of rules, which players must follow, which are illustrated in the sidebar on the following page: 'VUCA Paradigm of the 21st Century'. The game often has boundaries such as a racquetball court, baseball park, or supply chain. The game also requires specific skills to compete. A professional baseball player runs bases; hits a curveball, fastball, or slider; and fields a ball. Players keep score. The game score defines winners and losers.

CONTEXT: VUCA Paradigm of the 21st Century

- Understand what it means to live in VUCA time in a VUCA world.
- Understand the 'new normal' in your business sector and manage appropriately.
- Do not play by the dominant competitive or commonly accepted rules of an industry or sector.
- Compete based on new online business models and platforms, not only on processes, products, and services (RBDM).
- Globalize the end-product manufacturer's perspective and knowledge base.
- Develop Risk Based Problem Solving (RBPS) and Risk Based Decision Making (RBDM) capabilities to compete effectively.
- Be proactive, preventive, predictive, and preventive in SCRM.
- Get innovative or shortly become dead.
- Break down customer-supply barriers.
- Use people's knowledge and skills at the right time.
- Turn competitive learning into a corporate religion.
- Develop Key Performance Indicators (KPI's) and Key Risk Indicators (KRI's) to measure supply risk performance.

Supply and operations managers more often perceive their world through under-performing suppliers and broken links in global supply chains. Or in other words, operations managers perceive their world through the VUCA lens of Supply Chain Risk Management (SCRM).

In much the same way, SCRM is a game with risk principles, frameworks, processes, practices, and tools, which we discuss throughout this book. VUCA is also driving a shift from 'SCRM Old School' to 'SCRM New School' management. The change in each VUCA element is illustrated in a sidebar on the following pages.

PARADIGM SHIFT AS RISK DISRUPTION

A paradigm shift is a sudden and dramatic change in game rules. Paradigm shifts can be monumental or small. Usually, small shifts serve as a precursor to larger changes. What was right before may now be wrong.

What was the pathway to success may no longer be the case. For example, supply requirements seemingly acceptable five years ago are no longer accepted such as prison labor and lack of cyber security.

Paradigm shifts are difficult for managers who were hired, taught, recognized, promoted, and reinforced for a set of behaviors and activities that are now either unacceptable or have radically changed. This has happened in many traditional purchasing, materials management, design, manufacturing, quality, and planning functions with end-product manufacturers and their supply bases.

A paradigm shift was thought to occur in select areas or functions once in a lifetime. Now, a paradigm shift is so dramatic and disruptive it has become the norm instead of the rare occurrence. It is happening in business, regulations, professions, and technology. One way to manage supply change is to understand Supply Chain Risk Management rules and boundary conditions.

Even in this new supply chain game, there are some fixed rules, fundamentals, and boundaries. No matter how many balls, strikes, or bases, players still have to run, throw, catch, and swing. And even in a game with floating rules, some end-product manufacturers can create winning competitive strategies focused on the SCRM guidelines covered in this book.

SUPPLY CHAIN RISK MANAGEMENT PARADIGM SHIFT

The evolution from purchasing to supply management to SCRM has been a paradigm shift. As discussed, paradigm shifts occur when the fundamentals and assumptions of a process, discipline, or business model change. We can say the supply management function as we knew it five years ago has seen a paradigm shift to SCRM.

SCRM is sometimes described in terms of the following set of changing rules, boundary conditions, tools, and expectations:

- **Set of rules.** Most written and implicit SCRM rules have changed in the last five years. Much of this is due to changes in technology and the rise of ecommerce. Volatility, Uncertainty, Complexity, and Ambiguity (VUCA) are driving the new set of SCRM rules involving

SCRM Volatility – New School Solutions	
SCRM Old School	**SCRM New School**
Is inwardly focused	Is outwardly focused
Accepts the status quo	Thinks about tomorrow's risks before they occur
Expects cause and effect relationships	Understands risk correlation is more common than causality
Process stability	Process innovation

1. Risk Based Problem Solving (RBPS) and 2. Risk Based Decision Making (RBDM).

- **Boundary conditions.** A few years ago, the boundary conditions for almost all supply management involved the end-product manufacturer and final-customer. Now, SCRM boundary conditions involve the entire supply value chain of end-product manufacturers, final-customers, and even interested parties. We are seeing this broadened boundary with ISO 9001:2015 requiring companies to consider 'interested parties' in the scope of the ISO certification.

- **Tools.** The tools of the operations manager ten years ago were faxes, cell phones, laptop computers linked via e-commerce intranets, enterprise resource planning, and the Internet. Now, the tools include cloud computing, artificial intelligence, IoT, 3D printing, cyber security, and predictive tools. ISO 9001:2015 now requires the demonstration of Risk Based Thinking, which include RBPS and RBDM tools.

- **Expectations.** A manager entering the workplace ten years ago assumed lifetime loyalty was rewarded with lifetime employment. In much the same way, suppliers entering into a business relationship with the end-product manufacturer expected long-term, reliable margins. No longer! Like employees, suppliers expect their contract to be in place as long as they offer value-adding capabilities and processes. Traditional supply loyalty to the end-product

SCRM Uncertainty – New School Solutions	
SCRM Old School	**SCRM New School**
Clings to existing processes when they are unstable	Develops new stable and capable transforming processes
Relies of past solutions to solve today's problems	Recognizes opportunities to change processes and seeks to adapt (RBPS)
Wants known business objectives and plan to achieve them	Understands that objectives are fluid and flexibility is critical for success
'This too shall pass' attitude	Anticipates uncertainty to develop new or modify existing business models[41]

manufacturer was rewarded with a long-term contract. Now rewards are based on cost, quality, delivery, and technology innovation or managing upside and downside risks.

VUCA PRINCIPLES

Most of the companies that rated among the top 100 today were not there 20 or even 10 years ago and some had not even started.
Anonymous

VUCA leads to disruption, which results in new rules for competition and SCRM. Supply chains are redesigned based on new business models and technology. For example, online retailing and sourcing are now dominant. SCRM is one critical result or solution implemented as a result of disruption.

Operations managers if they are to be effective must know the rules of the disruption game. End-product manufacturers, employers, internal customers, and suppliers base their business decisions on competitiveness, business model, platform, and new sourcing rules, which more often are disruption rules. What do we mean? Several supply examples may illustrate

SCRM Complexity – New School Solutions	
SCRM Old School	**SCRM New School**
Looks at an issue but ignores the context and environment surrounding the issue	Considers the context, issues, stakeholders, and environmental relationships surrounding an issue
Attempts to understand the entire issue before acting	Defines what is in and out of scope and moves forward
Is overwhelmed by complexity	Simplifies as much as possible complexities surrounding an issue

the point. Autonomous trucks are disrupting logistics. Drones are disrupting small package delivery services. Internet of Things (IoT) products and devices are able to communicate with each other (RBDM).

SCRM PRINCIPLES

There is no commonly accepted list of VUCA principles and SCRM assumptions. It seems every supply chain guru and consultant has his or her own set of beliefs. The following are some generally accepted SCRM principles that are discussed and elaborated throughout this book:

- **VUCA time.** We live in VUCA time. The result is the primary filter for most life and work problem solving (RBPS) and decision making (RBDM) is risk.

- **VUCA competition is global and fierce.** This decade will focus on global competitiveness. SCRM is a critical means to position an end-product manufacturer, its products, and services for continued profitability.

- **SCRM is a strategic business issue.** SCRM matures to become part of an organization's critical mission as an end-product manufacturer determines its product mix based on blended internal capabilities and supply-partner core capabilities.

SCRM Ambiguity – New School Solutions	
SCRM Old School	**SCRM New School**
Is uncomfortable with ambiguity	Uses ambiguity to act, innovate, compete, and monetize
Seeks structure and direction	Is comfortable with fluid structures and movement to solve a problem (RBPS)
Cannot identify the right problem to solve and starting point of the problem	Has a feeling for critical issues, frames them, and creates 'what if' scenarios

- **Competitively priced, high-value products and services sell.** Studies assert that real and perceived value in brands, products, and services sell. In industrial, commercial, and consumer buy decisions, value at risk is either the top criterion for making the buy decision or near the top along with cost, delivery, and service (RBDM).

- **Need for competitiveness, value, and innovation drives all business decisions, actions, and product development.** Competitiveness is part of the supply chain's vision, mission, and plans. VUCA and disruption drive the need for change and product innovation.

- **Anything that diminishes value is a risk.** Value creation (upside risk) is enhanced and value detraction (downside risk) are risk-controlled and mitigated.

- **Globalization and offshoring are redefined.** Global sourcing is reevaluated as each country is now focusing on internal markets and domestic sourcing, such as 'Made in Japan,' 'Made in India,' and 'Made in the USA.'

- **Final-customers are getting smarter.** Final-customers will shop for the best buy, will pay for only what they really want, and will switch products or service suppliers in a heartbeat. What does this mean for the supply chain? More change and uncertainty create opportunities to distinguish a supply chain from its competition. Uncertainty also provides opportunities to develop new products and services and even develop disruptive business models.

- **Sales produce profits but not necessarily jobs.** If high-value products and services sell at sufficient profit margins, they should generate continued profits for the end-product manufacturer. However, profits do not always ensure jobs as companies outsource critical jobs.

- **Stakeholder satisfaction produces sales.** The focus on final-customer satisfaction now includes the acknowledgement and satisfaction of different stakeholders and interested parties. Final-customer satisfaction and value production are the goals of all stakeholders including management, end-product manufacturers, suppliers, employees, stockholders, community, and the supply chain. Final-customer satisfaction means every customer experience is positive and pleasing so the result is a repeat purchasing customer. Final-customer satisfaction is achieved through managing controllable value factors, such as managing supply cycle time, quality, communications, risk, technology, and performance.

- **SCRM stakeholders are empowered to identify and control risks.** Responsibility and authority for SCRM activities are pushed down to the lowest organizational level and to the appropriate suppliers.

- **Final-customer satisfaction is facilitated through SCRM.** SCRM focuses all stakeholders and interested parties on satisfying final-customer requirements. For example, automakers can offer color and trim options almost on-demand.

- **'Make or buy' decision.** The 'make or buy' decision is the first risk based, sourcing decision (RBDM) an end-product manufacturer makes. If an end-product manufacturer has core process capabilities or core competencies then it would make the product. If it does not, then it buys (sources) the product from 'world-class' suppliers with complementary core process competencies. Now the 'make or buy' decision is more complicated due to offshore price increases and domestic content politics.

- **Partnering with risk appropriate suppliers.** The outsourcing decision goes something like this. If an end-product manufacturer understands a supplier has the same business model and ethic as it does then both can synergize their distinct core competencies (RBDM).

- **Sourcing function adopts new technologies.** Cycle time, JIT, lean, quality, IoT, robotics, benchmarking, 3D printing, artificial intelligence, cyber security, and logistics management are tools to facilitate SCRM. Basic to all SCRM activities is the need to continuously innovate and use the appropriate technologies to be competitive. What was good enough yesterday is not sufficient for guaranteeing tomorrow's profitability. Continuous innovation is pursued throughout the supply chain. Suppliers must improve their capability and move up the SCRM maturity curve.

- **Risk-assurance, auditing, corrective action, and preventive action close the innovation loop.** Systems, processes, and products are audited as part of a robust risk-assurance program. If there are product nonconformances or deficiencies, corrective action eliminates their root-cause and risk management prevents their recurrence.

- **SCRM relies on self-management, risk-control, and self-initiative to root-cause fix problems.** SCRM requires supply chain stakeholders to identify problems, identify risks, intervene when necessary, strengthen risk-controls, deploy appropriate treatment, and eliminate the symptom and root-cause of the risk (RBPS).

- **Critical SCRM systems, process and product parameters are continuously measured.** To achieve performance benchmarks, key SCRM process variables are measured inside the end-product manufacturer and with key suppliers. End-product manufacturer requirements such as Key Performance Indicators (KPI's) and Key Risk Indicators (KRI's) are key benchmarks in the SCRM journey.

- **Waste and risk are eliminated.** A key SCRM objective is to become lean through eliminating all types of waste, which is another form of risk. Waste includes excess materials, inefficiencies, redundancies, poor service, duplication of effort, and high cost.

- **People are cross-trained and compensated for learning.** Continuous and breakthrough innovation requires continuous SCRM to keep up with changing final-customer and end-product manufacturer requirements, trends, and technology.

SUMMARY

Dun & Bradstreet, the global credit rating firm, recently said the following about SCRM:

> "Supply chain risk management helps companies identify and minimize threats that could interrupt access to goods or services vital to the business. Since many companies rely on vendors to provide raw materials, components, or finished products for sale, supply chain issues can cause severe financial hardship for the buyer. Many procurement professionals have adopted supply chain risk management strategies to enable them to anticipate and compensate for such threats."[42]

NEXT CHAPTER

Competitiveness is the driver for end-product manufacturers to 'focus on what they do best'. Supply Chain Risk Management is the means by which an end-product manufacturer can secure reliable suppliers that also 'focus on what they do best.' In the next chapter, we discuss SCRM and new business models.

CHAPTER 3:

BUSINESS MODELS

WHAT IS THE KEY IDEA IN THIS CHAPTER?

In a recent survey, supply chain risk was the #1 exposure to companies:

> "Not surprisingly, in a study of the 100 largest publicly-held manufacturing companies, consulting firm BDO discovered that the number one risk for these companies is supply chain disruption, specifically involving suppliers, vendors, or distributors."[43]

A few years ago, there was an ad for Putnam Investments: "You think you understand the situation, but what you do not understand is that the situation just changed." Amen! When I think, I've got it, 'got' and 'it' both change. It is scary and at the same time exciting – the thrill of SCRM when the rules seem to change overnight.

SUPPLY CHAIN AS A SYSTEM

An airplane is a system of spare parts flying in close formation.
Orville Wright

Jack Welch, the CEO of General Electric, popularized the concept of the boundary less organization where artificial boundaries to resources, communication and cooperation are removed to improve business processes. Technology and instant communication facilitate the redesign of organizational processes through the removal of horizontal barriers within the end-product manufacturer, the vertical barriers in the organizational hierarchy, and the external barriers outside a department or business unit or in other words throughout the supply chain.

One of the powerful analytical tools in management theory is the systems approach, which explains complex relationships in simple terms. A system is a group of elements or processes that function together to achieve a common goal. Elements are parts of an organization, including suppliers, distributors, and other parties.

SUPPLY CHAIN IS A SYSTEM

This chapter presents a systems approach to SCRM where independent and interdependent supply elements and processes work together to deliver quality products and services that satisfy final-customer wants, needs, and expectations. These system elements are the critical elements of a company's business model.

Core processes generate value through innovation, uniqueness, and flexibility. The supply chain is a value chain based on an overall process from the smallest supplier up the tiers to the end-product manufacturer to the final-customer. And within each company, there are many sub-processes that link or feed the overall value chain.

SYSTEM OF SUPPLY CHAINS

The fulfillment of orders is really a series of process chains that operate simultaneously. At the broadest level, the supply chain consists of all the processes and activities to deliver a product from the field to the dinner table or from the mine to the work table. Throughout this process, risks are identified and mitigated.

Another way to look at the challenge is that the supply process chain includes all the activities involved in producing, storing, and delivering manufactured goods to their ultimate destination. It originates at the enterprise level, moves down to the programmatic/project level, and finally to the transaction/product level.

David Anders, Frank Britt, and Donavon in **The Seven Principles of Supply Chain Management** offered the following wisdom to supply managers:

> "These savvy (supply) managers recognized two important things. First, they think about the supply chain as a whole — all the links involved in managing the flow of products, services, and information

from their suppliers' suppliers to their customers' customers (that is channel customers, such as distributors and retailers).   Second, they pursue tangible outcomes – focused on revenue growth, asset utilization, and cost reduction."[44]

Value addition and elimination of value detractors/inhibitors are now critical for survival and long-term success. These are the success drivers of the supply chain.   Competitiveness, pleasing final-customers, and adding value drive all business decisions involving upside or opportunity risk (RBPS and RBDM).

So, today's rules of competitiveness come down to a few common-sense tips: know what you do best, focus on these core competencies, and out-source all other work. It sounds pretty simple but its execution is profoundly difficult. In **Blown to Bits**, Philip Evan and Thomas Wurster argue that every business will have to pick apart its processes that lie behind its products and put their value chain back together in different ways.

RISK BASED PROBLEM SOLVING AND DECISION MAKING

While this sounds simple, there are profound risk based problems and decisions that must be confronted. For example, the scope of SCRM often depends on the context and perspective of the user. Does the supply chain include all suppliers of all products and services throughout the product lifecycle from final-customer needs to final delivery? Or, does the supply chain include only critical suppliers one level deep in each supply chain (RBDM)?

Also, the integrated supply chain may well extend beyond first-tier suppliers to include sub-tier suppliers, transportation, warehousing, manufacturing, engineering, and final-customers. This makes it difficult to manage so the chain is often defined or scoped to the 'critical few' supply–partners.

This is a critical issue in terms of scoping the SCRM initiative. If the scope of the SCRM includes all suppliers, then an end-product manufacturer will find itself trying to boil the ocean – doing too many things at the same time. This is often a sure recipe for SCRM disaster. We have many hard lessons learned with SCRM and ERM, which we try to share in this book. We use the expression 'boil the ocean' as a reminder not to do certain things.

In order to not boil the ocean, end-product manufacturers often limit a SCRM program to critical first-tier suppliers. Once these are on board the SCRM program, then additional first-tier supply risk-controls can be designed and deployed down the supply chain.

In the following sections, we discuss several sourcing business models that will help you define the initial scope of your SCRM initiative.

CORE PROCESS BUSINESS MODEL

Successful big corporations should devolve into becoming 'confederations of entrepreneurs.
Norman McCrae, Editor

Each company has a set of core business model competencies or capabilities, in other words something that the company is exceptionally good at. The company can use its core capabilities to distinguish itself to final-customers or to differentiate itself from competitors. For example, Amazon has an online purchasing model, based on low-price and rapid delivery of product. Nordstrom offers exceptional final-customer service in the apparel trades. In VUCA time, SCRM is a core differentiator for organizational competitiveness.

CORE COMPETENCIES

Now, most business models focus on core process competencies. Why? In today's global economy, individual customers want customized products and services, not me-too global products. Thus, we see the rise of mass customization strategies to satisfy different customers based on a standard product chassis on a truck and customized peripheral elements, such as product color or after-market add-ons. To produce and sell products in small quantities in niche markets requires integrated supply chain processes. The result is that in the last few years we have seen the development of the core process business model.

Several years ago, the *Economist Magazine* summed up the core process business model neatly (illustrated in figure 5):

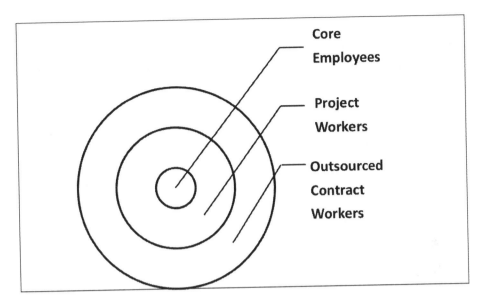

*Figure 5: **Handy Work Model***

"Focusing on core competencies might mean outsourcing peripheral employees, but it also means creating a core group of loyal
employees who have worked for the company for years; who have
absorbed or helped to create its ethos; and who are committed
enough to transmit that ethos to future employees. ... The focus is
increasingly on finding non-financial carrots to retain these 'core
employees' (if the only reason an employee stays with any particular firm is money, then he or she is much easier to poach."[45]

HANDY WORK MODEL

Charles Handy, probably today's smartest workplace guru, said more than
twenty-five years ago that work is changing. He developed a visual model
of work based on three concentric circles (see Figure 5), which we expanded upon:

- **Inner ring.** The inner ring consists of the organization's core competencies that distinguish it from the competition. The core process
  owners are in charge of the organizational future – keeping an eye
  on the competition, on new markets, and on strategy. The center
  is also in charge of the organization's overall architecture, specifically how it does work.[46] The inner ring is composed of corporate

executives, insiders, operations managers, and professionals. They are the glue that holds the organization together and grows it. These insiders are highly trained executives, professionals, marketing strategists, design engineers, supply managers, operations managers, and accountants who sustain the institutional memory. These full-time employees define the organization's vision, mission, principles, culture, and ethics.

- **Middle ring.** Middle ring is composed of project workers who support the organizational core processes. Project workers are full time employees, contractors, or temporary workers. We call these project people because they are mainly involved in discrete projects with a definite beginning and end. These people provide special skills, knowledge, and abilities that add organizational value.[47]

- **Outer ring.** Outer ring work is composed of largely interchangeable contract workers. These workers are often less-skilled service employees. Many are marginal workers who service the repetitive needs of the organization such as food service, administrative chores, or travel services.[48]

Does this work model reflect how supply chains are organized? We believe this simple illustration demonstrates the rationale for most if not all outsourcing. For example, an end-product manufacturer may want to develop a complex new product. The end-product manufacturer does not have internal resources so it will complement its core team with contractor-specialists or with key supply-partners with complementary skills. For example if a fast-food restaurant anticipates a spring or summer rush of business, it will hire temporary workers to ensure there are sufficient people to prepare food and service customers.

THE VIRTUAL BUSINESS MODEL

Manage the opportunities change offers.
Advertisement.

Some end-product manufacturers are going to extremes to focus on core competencies by 'virtualizing' most or all of their businesses. Remember when companies made things? More companies are moving to a virtual business model with few hard assets. Their core assets will be intellectual capital: trademarks, brands, people, inspiration, knowledge, and a SCRM business model. This is already practiced in software, consulting, and legal services. A company's work will consist of core process workers and special project workers who feed the core processes. This cannibalization or hollowing out of a business is sometimes called 'brand management.'

This virtual business model is predicated on an effective and efficient supply chain. Some end-product manufacturers are transforming their supply base into a virtual supply chain enabled by the Internet. The key driving factors for this include the need to decrease cycle times, increase operating flexibility, and reduce total supply chain costs. To achieve this, critical supply chain partners must identify and focus on their core competencies. Let us look at some examples of virtualizing business:

GHOST CARS – GHOST BRANDS

Forbes Magazine ran a cover story several years ago on 'Ghost Cars, Ghost Brands' featuring Bob Lutz, the former Vice Chairman of Chrysler. He developed an auto company called Cunningham Motor Company that he predicted will have $100 million of annual revenues – all with less than 20 employees. "Cunningham will be the world's most virtually integrated car company."[49]

HILLSHIRE BRANDS (SARA LEE)

Sara Lee part of Hillshire Brands sold its non-core factories. It decided to focus on its core strengths, specifically developing new products, managing its brands, and increasing market share. Sara Lee outsourced commodity manufacturing and other non-core activities and only retained its 'highly proprietary' processes. In other words, it focused on 'what it did best.'[50]

BRITISH AIRWAYS

Even airlines are becoming virtual because of competition and deregulation. British Airways to cut costs and increase operational flexibility leases its air fleet by the month, leases aircraft engines by the hour, and hires outside pilots and mechanics.

The airline is looking to suppliers to provide everything from spare parts to routine maintenance to flight training. The airline envisions it will supply little more than its brand name and its schedule. This could be a problem with regulators who worry about safety and from passengers who put their trust in big-name airplane brands.[51]

AMAZON.COM

Amazon.com reinvents itself and changes its business model almost yearly. Amazon.com is the poster company of the 'New Economy' and for ebusiness. Amazon.com started as an online bookseller but has morphed into a purveyor of new goods and services. In its latest incarnation, its core business is to collect final-customer buying information and then pull this information together into individual customer profiles. These are used to promote and sell unrelated products such as toys, electronics, books, and videos to specific customers. This type of virtual business morphs based on market conditions and final-customer requirements. Its core business is no longer a product or service, but an innovative and ever-changing business model.[52]

RETAIL DIGITIZATION

As more end-product manufacturers are evolving their online channels, brick and mortar stores sales are plummeting. Digitization is forcing all end-product manufacturers to reevaluate their business models and strategies.

The Amazon effect is the growing trend of many consumers to purchase online through Amazon, eBay, or similar retail platforms. Increasingly end-product manufacturers want to control the delivery process from production to consumer to maximize profit margins and to mitigate risks by bypassing retail stores or traditional distribution centers and selling directly to consumers. This challenges traditional brick-and-mortar retail stores who must now compete with online sellers and direct manufacturers who are struggling both to maintain sales volume and maintain margins.

THE AUTO INDUSTRY

The auto industry already outsources up to 80% of its manufacturing to suppliers, but now we are seeing auto companies outsource most of their non-core processes. Some auto manufacturers are even outsourcing their core work so they can focus on their brand management.

OUTSOURCING BUSINESS MODEL

During the last war, eighty percent of our problems were of a logistical nature.
Field Marshall Montgomery

Several years ago, I was asked to distill the rules of global business. I facetiously called them: 'Hutchins's Rules of Global Business.' I think they are still applicable. The major points were:

- Please your final-customers and interested parties.

- Identify what you do best.

- Institutionalize core processes.

- Outsource non-core work to 'world-class' suppliers.

- Acquire supply business model, processes, and products.

- Measure supply performance.

- Innovate and improve continuously.[53]

It is interesting that each of the above steps involved some form of RBPS and RBDM. Let us explore each one:

Please Your Final-Customers And Interested Parties

Since Tom Peters wrote his opus In **Search of Excellence**, pleasing the customer has been a conventional wisdom of business excellence. The customer was first considered the external customer, the final user; then the concept was enlarged to include the internal customer, the person next in line adding product value or the end-product manufacturer. Then the supply chain definition expanded to include all customer-supply links.

The new ISO 9001:2015 standard impacts more than 1.2 million companies. ISO 9001:2015 certified companies must consider the requirements of stakeholders, end-product manufacturers, final-customers, and interested parties. The end-product manufacturer is only one of a growing group of external and interested party supply chain stakeholders that must be satisfied in a global economy.

A customer may also be part of the external supply chain links, which may include local authorities, state government, or federal government. More and more, regulatory authorities, dealing with safety, environmental, consumer and health issues, are growing in importance. Therefore, a partial list of SCRM stakeholders would include final and internal customers, management and employees, distributors, shareholders, government, suppliers, and unions.

Satisfying if not pleasing these stakeholders becomes more important. It is said that any of these stakeholders may have veto power for a company to enter new markets, be profitable, and even survive.

Identify What You Do Best

A company cannot be all things to all people. An end-product manufacturer must focus on what it does best. In a competitive economy, the goal is to become 'world-class' - the best in a market in one or several key processes. Best may involve designing widgets, supplying typing services, providing legal work, assembling printed circuit boards, or sourcing products. Best requires that a company discovers and focuses on its core process competencies. To become the best in Supply Chain Risk Management requires simplicity, lean thinking/doing, risk management, and control of the supply chain.

As large end-product manufacturers attempted to please final-customers with a large variety of products, several problems arose. Products would have different features, performance, or external aesthetics. Large plants would be built with parallel management, design, production, marketing, and distribution systems. Overlapping systems created increased variation, miscommunication, unbalanced flows, system constraints, redundancies and other risks that inevitably resulted in confusion. Bureaucracies were built that tended to protect turf instead of pleasing the final-customer. All of which are anathema to a lean, risk managed, and integrated supply chain (RBPS).

As well, no company ultimately has the ability to be the best in each business area. Being the best requires an abundance of resources that no organization has. What often occurs is that resources are spread too thin and the organization does many things only moderately well.

Institutionalize Core Processes

Core processes are essential to the operation and success of any business. Core processes are often horizontal spanning different functions, plants, and departments throughout the end-product manufacturer. For example, supply chain core competencies may involve engineering that develops robust drawings and accurate bills of material; involve production that has special handling, delivery, or packaging abilities; involve quality that conducts special testing and supplier audits; and involve accounting that ensures timely accounts payable and accounts receivable. The supply management function links these disparate activities into an integrated core process.

Often a company may only have one strategic area of excellence, internal capability, or core competency, which may involve SCRM, low-costs, state-of–the-art research and development, critical management abilities, special equipment, quality culture, team effectiveness, technology/knowledge, or distribution strengths.[54] More end-product manufacturers now view SCRM as a core competency to be developed.

For years, companies made a business case for focusing on core process competencies. J.P. Morgan, the investment bank, once concluded that American companies that focused on the one thing they did well outperformed the stock market by 11% while diversified companies underperformed the market by 4%.[55]

Outsource Non-Core Work To 'World-Class' Suppliers

The rationale for outsourcing goes something like this. We, the manufacturer, are in the business of making widgets so why should we spend our time focusing on running our own telephone company, information technology department, or training organization. The company will spend its time on things that will make a real difference to the final-customer, impact the bottom-line, and leverage its core competencies.[56]

The process of outsourcing non-core work is called brand management, virtual work, disintegration, or strategic sourcing. The result is the same. End-product manufacturers focus on core activities and outsource all other work. This has occurred in a major way over the last ten years as globalization continues.

Outsourcing systems and subassemblies to a capable supply chain can be a major differentiator and value add to an end-product manufacturer for the following reasons. First-tier suppliers assume responsibility for meeting specifications on a turnkey basis. Sometimes, second-tier or lower level suppliers are non-union shops where the costs of rework and scrap are typically lower than in-house assembly workers. Finally, outsourcing the subassembly makes for a cleaner and less cluttered assembly space where the subassembly is bolted on or fixed onto the final products. This helps reduce in-house cycle times.

Customer-supply partnering is the relationship between the end-product manufacturer and supplier. In some end-product manufacturers, vertical integration is losing its appeal to new forms of cooperation, coordination, and communication involving supply-partners. Global competition, high product development costs, high-quality expectations, low-cost requirements, shortened product life cycles, and individual customer requirements are accelerating the change to new forms of customer-supply integration and partnering.

Supply-partners are expected to be very good at what they do; in other words, to be 'world-class.' More often, work is sourced to a single supply-partner and/or acceptable alternate supplier for each product line. The rationale for selecting two suppliers is that one supplier may not change quickly, continuously innovate, and continuously lower costs. As well, the end-product manufacturer assumes inordinate risk with one supplier in terms of an act of God, strike, or other unforeseeable event stopping shipments.

Supply-partners are often first-tier partners, that are nimble entrepreneurs that know local customer requirements, have access to global information,

---

CONTEXT: Business Model Questions

- What are the key SCRM questions facing the end-product manufacturer?
- What information needs to be found to answer the above question?
- What was the company's current business model?
- What is the company's proposed business model to compete in VUCA time?
- What does the company outsource and keep internally?
- If the company has to outsource, can it be sourced domestically?
- What domestic suppliers can be identified to do what needs to be done?
- How does the company onboard suppliers quickly so they understand requirements?
- What metrics/objectives will the company use to manage these suppliers?
- How does the company set up a SCRM program quickly?

---

and are monomaniacal in the pursuit of pleasing the final-customer. Flexibility, quickness, and agility are the code words of their success.

This is happening worldwide. Many Japanese firms form keiretsus, which are formal financial, engineering, manufacturing, and supply networks. Japanese transplants are importing this model to the U.S. While still new in the U.S., these relationships are characterized by close product development partnerships. Cost, design, delivery, customer service, and proprietary information are commonly shared. The hoped-for results include: minimized process variation, eliminated redundancies, reduced supply risks, enhanced communications, and improved co-ordination.

Acquire Supply Business Models, Processes, And Products

Many supply chain management professionals still focus on the transaction - buying products instead of securing supply processes and even business model systems. This requires a shift of thinking, moving away from simply buying products to securing a reliable and lower risk source with robust, stable, and capable SCRM processes. If the means by which a product or service is produced is reliable and risk-controlled, then the outcome should

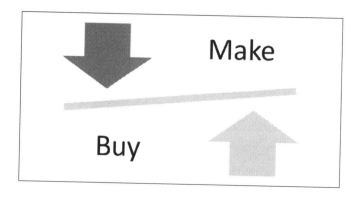

Figure 6: *'Make' or 'Buy' Decision*

meet contractual requirements, surpass final-customer expectations, add value, and ultimately secure revenue.

Modern SCRM assumes if final-customer and end-product manufacturer requirements are understood and internalized, then supply-partners have the risk-control systems and processes in place to address these requirements. If these processes are risk-controlled and improved, then there is a high level of assurance the products or services coming out of these processes satisfy end-product manufacturer expectations.

Measure Supply Performance
How is 'world-class' determined? Supply performance throughout the contract or product lifecycle is continuously monitored and risk-assured. Traditionally, commercial buy decisions were based on price, availability, technology, and delivery. Similarly, consumer buy decisions were largely based on price, branding, and packaging. Now industrial and commercial buy decisions are more complex, based on verifiable risk attributes, quality, total cost, eye-catching design, environmental friendliness, compliance, and other factors, which will be discussed in the next chapter (RBDM).

While consumer decisions are often the result of on–the-spot visceral responses, most industrial risk and commercial buy decisions are based on well thought out, researched SCRM return on investment (ROI).

Innovate And Improve Continuously
Standing still in a fast-moving economic stream is the equivalent of moving backwards. Treading water is death especially in technology markets. The obvious solution is to innovate and improve continuously. Innovation is the ability to conceptualize and commercialize new products. Improve is the ability to control and minimize supply chain variation so there are fewer defective products and less overall risk.

Theoretically if a company builds on its strengths, it will consistently achieve competitive returns from its business. Final-customers will flock to it because of its distinguishing characteristics, products, or services.

A cohesive mission, strategies, and objectives allow business units, suppliers, plants, teams, and processes to be linked and synergized. The challenge for the operations manager is to develop, nurture, and reinforce these unique characteristics throughout the supply chain.

'MAKE OR BUY' RISK DECISION MODEL

Any business must always plan ahead, either to capitalize on success or to reverse the trend if not successful.
Anonymous

One of the first critical risk decisions is to determine the manufacturing business model for a category of products or services. One business model is to make products internally based on just-in-time of the placement of the order. Another business manufacturing model is to outsource products or services to suppliers. This is called the 'make or buy decision.' Sometimes, this is an easy risk based decision (RBDM) or can be very difficult.

'MAKE OR BUY'
The 'make or buy' decision is one of the most critical supply chain, risk based decisions (RBDM). The SCRM organization has a key role in this decision. The decision is important for several reasons. It determines and defines an end-product manufacturer's core competencies. It determines what level of investment the business will make internally as well as with suppliers.

CONTEXT: McKinsey Digitization Competition Model

- **New pressure on prices and margins.** Online customers can compare product and service pricing and performance with a few clicks and if necessary, switch companies within seconds.
- **Competitors emerge from unexpected places.** Barriers to entry and product differentiation disappear as companies can use Amazon, eBay, and other platforms to warehouse and distribute products. They can use PayPal or the platform's own revenue system.
- **Winner takes all dynamics.** Online companies with the overall best value proposition wins.
- **Plug and play business models.** Digitization reduces transaction costs. The result is that supply chains often disaggregate. Broken supply chain links and gaps are filled with online startups and disruptive competitors.
- **Growing talent mismatches.** Software, artificial intelligence, and automated online processes replace people including experts.
- **Converging global supply and demand.** In many markets, suppliers are now offering products and services that compete with and/or complement the end-product manufacturer's or service provider's.
- **Relentlessly evolving business models.** Digitization is one of many VUCA drivers to change business models. Think IoT, 3D printing, autonomous vehicles, and the list goes on.[57]

The 'make or buy' decision involves financial and capability issues as companies ask: 'Do we have the expertise to manufacture a quality product and deliver it at a competitive cost?' Since some industrial tasks cannot be effectively accomplished in-house because of lack of equipment, trained personnel, or material, the answer to the question is often 'no.' So, non-core products and services are contracted to outside suppliers. The decision to outsource is usually based on real business reasons (RBDM):

- End-product manufacturer decides to focus on core competencies.

- End-product manufacturer wants to develop strategic relationships and customer-supply partnerships in emerging markets.

- End-product manufacturer wants access to the supplier's marketing, innovation, skills, and technology.

- End-product manufacturer anticipates substantial cost savings from outsourcing.

- End-product manufacturer gains access emerging markets.

'MAKE OR BUY' IN TECH COMPANIES

Let us look at a high-tech company's 'make or buy' decision-making. Following the rule, 'cannot be all things to all customers,' high-tech companies focus their internal resources on core technology while depending on strategically outsourced innovations to complement their internal efforts.[58]

In general, high-tech companies such as Intel and Microsoft competitively position themselves based on their core knowledge competencies so that internal development ('make' decision) provides the most competitive advantage. In areas away from their chip design and software development core competencies, they may outsource, license, or purchase required competencies (RBDM).

GUCCI MAKE OR BUY DECISION

Gucci on the other hand decided to expand its internal capabilities and competencies to assure its quality reputation through internal design, manufacturing, and distribution.

Gucci sells expensive products, such as multi-thousand dollar handbags, shoes, and leather goods. Its business model traditionally was low volume - very high margin. The Gucci label was worth a lot of money due to its brand, exclusivity, and design. Design was always paramount in the buyer's decision. The challenge for Gucci is demand for its products is surging. The question is how do they meet demand? The supply chain cannot keep up with demand. So, it is revamping and redesigning its supply chain model to make products internally and to integrate vertically.

Insourcing or internal manufacturing is so important to Gucci that it reported:

"Gucci's new production strategy, taking a step towards enhanced vertical integration and reduced lead time in a fashion world where supply chain is becoming as important as design."[59]

'MAKE OR BUY' DUE DILIGENCE

If a company outsources a product or service, then it can work with existing suppliers or find new suppliers. As much as possible, companies do not want risks, surprises, waste, or unwelcome uncertainty. They want consistency that can be gained through Risk Based Problem Solving (RBPS) and Risk Based Decision Making (RBDM). They want to work with known people, known relationships, and known supplier processes. Another way to say this is that end-product manufacturers want to minimize downside risks and optimize upside risks. It is pretty simple; life and business work better when we work with knowns. Again, think variability and risk. We do not want unknown variability, unknown risks, unknown people, unknown processes, unknown suppliers, or unknown outcomes.

One solution is to encourage supply-partnering relationships. End-product manufacturers and suppliers trust each other to share key process information, technologies, cost/delivery/quality targets, and even investments. This frankly is not easy. It requires deep trust that a nondisclosure agreement (NDA) cannot enforce.

The 'make' decision also is not easy for a supplier. The supplier may even pass on the opportunity to provide the product or service because of unknown risks. The products may not be worthwhile to manufacture or deliver the requested services. The products are low volume or 'one of a kind' that may require new production equipment or provide insufficient margins. Is the end-product manufacturer willing to pay for the added supply investment? Many questions - few easy answers. The 'make or buy' decision usually comes down to optimizing many risk related factors (RBDM).

RISK REWARD DECISION

Also, the 'make or buy' decision involves a 'risk/reward' or 'cost/benefit' analysis. For example, low value products are usually commodity and non-strategic items. As well, there are multiple suppliers who can produce this commodity so the risk of losing a commodity source or finding competitive bidders is relatively low. If the supplier provides a high-value, innovative product, or process technology, the end-product manufacturer may partner with a supplier or bring the product in-house (RBDM).

What does a company do if a new or existent supplier cannot produce the product to the end-product manufacturer's requirements? The end-product manufacturer has several options. It can find a new supplier or it can work with an existing supplier. The end-product manufacturer may even improve the supplier's capabilities. How? The end-product manufacturer can provide technical assistance, machines, incentives, or even pay the cost of improving the supplier's capabilities.

And, there is the 'risk-reward' decision of switching suppliers. This is not a negligible risk. The risk or cost of an unknown supplier may be too high. When should an end-product manufacturer change a supplier? The change will occur when the cost, pain, or risk of keeping the supplier exceed the cost of finding a new supplier.

All of these risk based strategies will only work if the end-product manufacturer and supplier have a mutual understanding of risks and possible rewards. The decision also depends on the 'risk-reward' profile of the product or service supplier. If there is not a mutual understanding of the risk-rewards, then the following occurs. Usually, the higher the product value, the higher the possible risk to the end-product manufacturer. The end-product manufacturer will require higher risk-assurance and risk-controls from the supplier, who may pass on the opportunity. The end-product manufacturer will then decide to in-source the product or service.

WHAT DOES 'MADE IN THE USA' MEAN?

Domestic sourcing for domestic consumption has become a critical national issue for many countries. We are now seeing 'Made in China', 'Made in Mexico', 'Made in India', and 'Made in the USA'. We are seeing similar developments in France, United Kingdom, Germany, Japan, Brazil, and China, where domestic sourcing is becoming a national jobs issue. This trend is called national in sourcing or in other words returning to the home country for design, manufacturing, distribution, and consumption.

President Trump of the United States is placing 'Made in the USA' requirements as a condition of procurement in military systems and high-tech

---

CONTEXT: 'Make or Buy' Critical Risk Questions

- What are the required technology and information core competencies?
- What are the risks and benefits of insourcing internally (make it), insourcing (buy it domestically), outsourcing to existing suppliers (buy it), or and outsourcing to new suppliers (buy it)
- Can internal processes make or service it cheaper, better, etc. than others?
- If the end-product manufacturer does not have internal processes to provide the product or service, what does the company have to do and how much will it cost to produce in–house?
- Who is the product or service competition?
- What are their core competencies and do they insource or outsource similar products?
- Do suppliers have core competencies that complement the final manufacturer's for providing the product or service?
- Which suppliers have competencies to make and deliver the product or service effectively, efficiently, and economically?
- Does the end-product manufacturer have the resources to manage, risk-assure, control, and mentor possible new suppliers?

---

products. The same is happening in China, India, and many industrialized countries. The result is that free and open trade is disappearing as countries set up trade barriers. Again, global trading risks increase. For example a complex product such as an automobile may have thousands of parts that are sourced globally. This begs the question: what does 'Made in the USA' really mean in terms of domestic content or complying with new rules-of-origin?

SUPPLY QUALITY MATURITY MODEL

Everything is connected to everything else.
Barry Commoner

Maintaining product and service quality continues to be the #1 challenge in SCRM. The concept of quality as process management and risk-control is thoroughly embedded in SCRM thinking and doing. Many mature supply chain initiatives employ lean, just-in-time, demand-pull, Six Sigma, quality function deployment, lifecycle costing, design for manufacturability and other mature process concepts. Quality, cost, delivery, risk, and technology are the major elements in most industrial and commercial buy decisions. Each element, such as cost or quality, has its own maturity curve. In this section, we will introduce the process maturity concept and apply it to SCRM (RBDM).

A recurring theme in this book is that SCRM's role is changing from procuring products at the lowest price to securing reliable, risk-controlled supply processes. This move from a low-price, product focus to a SCRM process emphasis is a paradigm shift. How does SCRM manage the product to process shift? Suppliers are asked or required to move up a customer-supply capability and maturity curve.

CUSTOMER-SUPPLY MATURITY

Supply management is responsible for assuring that supply quality, technology, delivery, cost, risk, and service processes can meet end-product manufacturer requirements and can complement internal core capabilities. This is the main reason SCRM is elevated to the executive level. This is critical to all end-product manufacturers that spend a large part of their manufacturing dollar buying products and services from suppliers.

What is the best method of assuring high-quality, low-cost products? A critical SCRM practice is to partner, develop, and integrate fewer suppliers into a seamless process chain. Suppliers often start as candidate suppliers then move up a maturity curve as their quality, cost, technology, design, delivery, and service process capabilities improve. At each step of the process, suppliers are induced, evaluated, and measured against demonstrable value attributes. This is called supply development. Critical suppliers are measured in each step of the journey against demonstrable criteria or standards, which may include ISO 9001:2015 certification.

Over the lifecycle of the relationship, suppliers mature and grow as supply-partners. Suppliers are expected to continuously improve their quality, cost, technology, and delivery process capabilities so the entire supply chain can maintain a competitive edge.

As SCRM increases in importance, more end-product manufacturers want to work with a 'critical few' or 'world-class' suppliers. What makes a 'world-class supplier?' This supplier has core competencies that the end-product manufacturer does not have. In other words, a 'world-class' supplier is one with known and demonstrable processes and is high on the quality capability and maturity curve shown in Figure 7 on the next page.

FROM PRODUCT INSPECTION TO OPERATIONAL EXCELLENCE

Leaders win through logistics. Vision, sure. Strategy, yes. But when you go to war, you need to have both toilet paper and bullets at the right place at the right time. In other words, you must win through superior logistics.
Tom Peters

Quality, cost, delivery, risk, and technology are the major elements in most industrial and commercial buy decisions. Each element, such as cost, risk, or quality, has its own maturity curve. In the supply process chain, quality still reigns as the most important factor in supply selection and development. In this section, we describe the quality capability and maturity journey.

The department responsible for quality activities has many titles these days. The department is called Quality Management, Quality Assurance, Quality Engineering, or Reliability Engineering. In the service sector, it is called Customer Service and Customer Quality. Regardless of the name, the departments have the same responsibilities, which are to coordinate, maintain, and monitor quality processes throughout the end-product manufacturer and into the supply base.

As the figure on the next page indicates, quality in most organizations matures along this path:

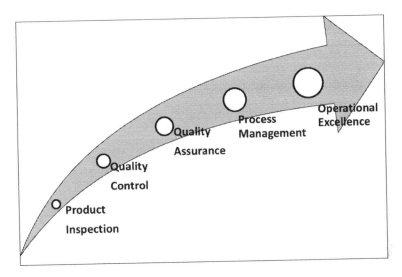

Figure 7: ***Quality Maturity Curve***

- **Product inspection.** Inspection is product based. The first quality function was an inspection organization whose main responsibility was to direct the work of factory product inspectors. Factory inspectors policed the quality of incoming, in-process, and outgoing products. The department had no independence because it reported to the head of manufacturing.

- **Quality control.** Quality control (QC) was the start of moving from a product to a process based approach where manufacturing processes and product variation was controlled. Why? Stabilized, in control production processes produce consistent and conforming products or in other words, control product risks. A department called quality control replaced the inspection department. As the name suggests, prevention through process control was emphasized. Quality control, now independent of manufacturing, monitored manufacturing performance. Quality could no longer be overruled or second-guessed by the department it was supposed to monitor. However, quality control still focused on manufacturing control and did not monitor engineering design.

- **Quality assurance.** As quality became more important, management realized it needed an independent and objective group to monitor internal manufacturing quality as well as quality in other functions, such as engineering, supply chain management, and distribution. So, a separate but equal department was created called quality assurance.

- **Process management.** As organizations matured, quality evolved into organizational process management. Quality is now prevention oriented and its scope has increased to involve the entire organization and the supply base. Process management or simply business management can involve activities including standardizing processes, decreasing process variation at required targets, improving time-to-market, becoming lean, increasing customer satisfaction, or improving process capability

- **Operational excellence.** At this level, the term 'quality' even disappears and becomes simply operational excellence. Operational excellence includes Six Sigma, lean management, risk management, and project management. Operational excellence also includes supplier's suppliers, second-tier and lower suppliers. Responsibility for work rests with the person, work unit, or supplier performing the work. At the simplest level, this is a person on the line or a person who delivers products to the customer. Only this person can add or degrade product or service value.

SCRM PROCESS MODEL

Reorganization is the permanent condition of a vigorous organization.
Roy Ash

The quality maturity model described in the previous sections is now replicated in customer-supply risk, design, cost, schedule, and technology management. As a product is made or a service is delivered, value is added. Value enhancing consists of adding labor, machinery, material, or facilities required to do work. On the other hand, value detracting is the loss of value or risk of not securing or deploying the labor, machinery, material, or facilities to do work.

CONTEXT: Supply Value Chain Assumptions

- All work is a process.
- Each process has inputs and outputs.
- Each process has a supplier(s) and customer (s).
- Each process is controlled, made consistent, and improved.
- Each process and each process step adds value.

UPSTREAM PRODUCT RISK MANAGEMENT

Value is ensured at the front end of the process or at the source, not at the end of the process. For example, it is expensive to scrap a porous casting after it has gone through 10 or more manufacturing steps. Or, it is counterproductive and risky to manufacture a highly reliable product if the delivery system is inefficient or customer experience is awful.

Customer satisfaction cannot be guaranteed by instructing a salesperson in courtesy after the sale. The salesperson or receptionist should have been instructed in policies and procedures before the assignment.

Likewise, quality as 'conformance to specifications' cannot be assured or risk-controlled by inspecting quality into the product. Inspectors inspect, sort, rework, rehandle, and reinspect defective products. This is wasteful, because it consumes time, expense, material, personnel, and facilities.

MOVING FROM A PRODUCT TO A RISK PROCESS ORIENTATION

The shift from an end-product focus to an upstream, process focus requires management and behavioral changes. For example, end-product manufacturers will purchase products from a candidate supplier based on acceptable product/service quality or delivery. These low-level relationships are usually based on low-price, any time delivery, and acceptable quality.

Low-level maturity purchasing is often transactional, at one point in time. Transactional relationships can deteriorate into adversarial, arms-length, win-lose, low-price, and short-term.

Process and risk based relationships are more mature and are mutually beneficial, life cycle cost based, and long-term. A process orientation implies the operations manager secures the supplier's core process competencies and the supplier's processes are aligned and integrated with the end-product manufacturer's core processes thus creating synergies (upside opportunity risk). The process of moving up the maturity curve is called supply development or customer-supply partnering.

SCRM AS SUPPLY PROCESS MANAGEMENT

SCRM is fundamentally a supply process management. Each process step from marketing to shipping a finished product should add value. Marketing determines what final-customers' needs are and communicates these to the end-product manufacturer. Engineering designs a product to final-customer requirements. Supply management buys components and parts to be machined or assembled into a finished product. Manufacturing produces the product. Distribution delivers it to the final-customer. Quality monitors conformance to requirements throughout the process. Suppliers provide products and/or services. This pattern is repeated each time a new product is developed or an existing product is modified.

In other words, all supply streams and chains are process flows. Examples of SCRM process flows include: information management, capital investment, product development, resource management, production flow, and product delivery. Each of these flows can vary, which can result in risks. There is usually an upper and/or lower limit of acceptable delivery, quality, cost, and other critical flow characteristics. If the variation is above or below the acceptable levels, then risks may exist.

Each critical step must demonstrate process control and capability. Control is the ability to be stable and consistent around a performance or risk target. Capability is the ability to meet a standard or specification consistently.

SCRM IS KEY

End-product manufacturers look to suppliers to risk-control and seamlessly integrate their processes into the supply chain. In other words, suppliers are expected to self-manage their quality, delivery, cost, and risk processes. This implies there is little incoming material inspection, products are delivered just-in-time to the proper location, and overall contract cost reductions are shared with the end-product manufacturer.

So, end-product manufacturers will increasingly audit suppliers, certify them, and expect a high level of performance. If suppliers comply, rewards are shared and they move up the maturity and capability curve. Suppliers are also induced by larger and longer-term contracts, technical assistance, and special equipment.

DOMESTIC SOURCING MODEL

One way to increase productivity is to do whatever we are doing now, but faster ... there is a second way. We can change the nature of the work we do, not how fast we do it.
Andrew Grove, CEO Intel

The outsourcing business model is used by end-product manufacturers that want to replicate and scale seamlessly and globally. The outsourcing and supply chain management phenomena have created significant challenges within organizations. The 'outsourcing' concept originally dealt with buying products from external suppliers. 'Outsourcing' now has become synonymous with people losing their jobs. The rationale can be understood through the Handy Work Model.

Let us look at Handy's work model from a sourcing point of view. Critical questions for the end-product manufacturer include:

- Do we want to outsource?

- If we do, do we outsource offshore or domestically?

As discussed, many end-product manufacturers source up to 80% of their manufacturing dollar to suppliers. However, these end-product manufacturers still are responsible for the quality and compliance of the supplier's products. What companies now realize is that as the proportion of the manufacturing value increases to the supply base; the end-product manufacturer has less control over second-tier and lower-level suppliers and encounters more inherent risks.

OUTSOURCING
Almost anything can be outsourced. The list of possible outsourced products and/or services include: IT services; project management; HR services; engineering; call centers; software development; and even executive management. A start up can even hire experienced senior managers to help the firm scale.

Outsourcing whether domestically or offshore is usually seen as a strategic decision. The rationale to outsource can vary. It is due to the more effective allocation of resources; being close to the market and final-customers; lower costs; or being close to academic centers of innovation

As outsourcing increases, the entire supply chain may become a single integrated, global manufacturing platform involving design, warehousing, logistics, and rapid delivery. Real-time analysis of supply chains and possible routes may not be able to identify or react quickly to potential bottlenecks, crises, changes in customer demand, inventory changes, and poor quality issues (RBDM).

While this is the hoped-for result, the reality is sometimes different. Global supply chains are very complex with many moving parts that that can and do fail. Inherent risks can increase immeasurably with outsourcing and particularly with offshoring. Hence, more consideration of domestic sourcing.

DOMESTIC SOURCING OR OFFSHORE SOURCING

Many end-product manufacturers source much of their manufacturing and non-core dollars to suppliers. In other words, this is where the money is as well as potential risks. Each supplier in the supply chain has its own business, operational, governance, business model, and operational risks to mitigate and intellectual property to protect.

The decision to insource is based on many factors such as the cost of transportation, closeness to the market, financial risks, and other factors. Domestic sourcing is sometimes called 'near shoring.' In other words, the end-product manufacturer will build design or production facilities near the final-customer. This is done for several reasons such as a retailer of high fashion wants to be close to its customers and make just-in-time clothing as opposed to waiting 6 months to a year to deliver the clothing to a customer from offshore suppliers (RBDM).

However, these end-product manufacturers and brand owners are still responsible for the quality and safety of the supplier's products. What end-product manufacturers realize is that as the proportion of the manufacturing

CONTEXT: McKinsey Decisions That Redefine Competition

Decision 1. Buy or sell businesses in your portfolio? Digitization as a VUCA disruptor can destroy growth and profitability of existing businesses. Portfolio of products, businesses, and business models may have to be altered to ensure continued growth and margins.

Decision 2. Lead your customers or follow them? Each segment of a business will be disrupted due to digitization. The question is whether to lead or follow the market. Tough decision that may require upside risk investments and acquisitions.

Decision 3. Cooperate or compete with new attackers? Local and global competitors are entering each business segment. Each with a new idea or variant of an existing business model. Each competitor can cause pain. Does a company decide to cooperate or compete with these new entrants?

Decision 4. Diversify or double down on digital initiatives? Where does a company invest or divest its resources to compete effectively? As with the other questions, it comes down to risk based decisions (RBPS and RBDM) – diversify risks and/or invest in one or two high probability outcomes and/or mitigate possible consequences and/or all of the previous.

Decision 5. Keep digital businesses separate or integrate them with current digital ones? Integration makes sense for synergizing resources. But, digital businesses require entrepreneurial mindsets, new business models, and actions that may not exist in more traditional management.

Decision 6. Delegate or own the digital agenda. The digital agenda requires executive management time, engagement, and technical capabilities. Should time and technical abilities be purchased, delegated, or outsourced? Again, more risk based decisions. [60]

value increases from the supplier base; the final manufacturer has less direct control of second-tier and lower-level suppliers. This problem increases exponentially as supply chains expand globally so political, environmental, and social events can disrupt the supply chain (RBPS).

KELLOGG DISTRIBUTION MODEL

Many end-product manufacturers are redesigning their business models and supply chains. For example, Kellogg is moving its U.S. snacks model business away from a direct store delivery to a warehouse model. This results in terminating distributor contracts. Kellogg expects to save 15% by redesigning the business model because 39 distribution centers and 4500 jobs are eliminated.[61]

Modifying or reinventing supply chain business models is painful for end-product manufacturers and their suppliers. They must change something that has worked well for years, adapt to new management models, and become innovative. Suppliers are also expected to change to accommodate the new business model.

Some end-product manufacturers are even redesigning themselves and acting more like startups. The hope is new business models will allow them to continuously improve, collaborate with suppliers, innovate with new products at faster rates, while improving margins. The challenge is that people need to learn new skills, use technology smarter, adapt and adopt new behaviors – all of which are difficult.

CROWD SOURCING FACTORY

G.E. opened a micro factory in the U.S. to create new product ideas as prototypes or in small batch runs. The idea is based on crowd sourcing, specifically reaching out to global online communities to rapidly introduce new products. Some of the new problems that G.E. has posed to its global community include: jet engine inspection, measuring reflective objects in hard to reach places, and compressing Computer Tomography (CT) medical scans so they can be stored easily and accessed quickly.[62]

SUMMARY
Over the last 20 years, offshoring has become the predominant supply chain business model. Much of this was driven by the globalization of markets and rising middle classes in developing countries.

However, the cost of offshore production for products is now weighed against the risks of sourcing in a specific country or region. Why? The cost of offshoring has increased as workers demand higher salaries. Reputational risks increase as safety and other issues become more critical. Quality concerns are increasing. Finally, Corporate Social Responsibility exposure and risks have increased. So, more end-product manufacturers are making the decision to source domestically, often using highly automated factories (RBDM).

The domestic or offshore decision becomes a risk based decision. The end-product manufacturer determines the risks and benefits of offshoring vs. domestic sourcing, and then makes a business decision what to pursue (RBDM).

NEXT CHAPTER
We have discussed competitive drivers, the need to focus on core competencies, and the benefits of outsourcing. In the next chapter, we discuss SCRM risk principles and values.

CHAPTER 4:
SCRM PRINCIPLES AND VALUES

WHAT IS THE KEY IDEA IN THIS CHAPTER?

SCRM in VUCA time is an emerging opportunity. Deloitte conducted a study of 600 C-level executives that revealed that 45% felt their SCRM programs are only somewhat effective or not effective at all. A mere 33% use risk management approaches to strategically manage supply chain risk.[63]

The key idea in this chapter is risk management is the equivalent of value management. Customer product, service, or experience value can diminish which results in consequence and likelihood risks. Or, customer product and experience value are enhanced through the proper architecture, design, deployment, and assurance of the specific value attributes.

SCRM VALUE PROPOSITION

The real issue is value, not price.
Robert Lindgren, Business person

As the above quotation says, the real issue in SCRM is value creation (opportunity) and risk mitigation (downside) along with value based pricing. SCRM sourcing is not an end, but a means to establish supply chain and operational value.

While an end-product manufacturer may perform well in many areas, following the competitiveness precept of 'stick to your core competencies,' end-product manufacturers will develop and focus on a few core differentiating strategies.

BUSINESS CASE FOR SUPPLY CHAIN RISK MANAGEMENT

Ian Stewart and David McCutcheon in **Manager's Guide to Supply Chain Management** made the case for strategic outsourcing:

> "The strategic objectives in outsourcing are relatively straightforward. Basically, firms are interested in how they can either significantly reduce product costs or add to what customers perceive as value-added benefits. Naturally, firms hope that the value-added benefits can be achieved at lower cost, but in such cases the cost role is subordinate. Value-added benefits might include improved delivery speed, additional design features and options, or the ability to be customized. Some of these benefits are best achieved by using in-house product design and process management capabilities."[64]

End-product manufacturers outsource because of 'value-added benefits' listed in the above quotation. 'Value-added benefits' include reduced cost, faster delivery, higher quality, better technology, or enhanced process management. Value-added benefits are mainly upside (opportunity) risks.

Risk to the end-product manufacturer or supplier is the lack of ability, hindrances, or obstacles to realize the 'value-added benefits.' One way to think about 'value-added benefit' is in terms of a business objective. If the end-product manufacturer or supplier cannot meet its business objective, then there is a risk. The goal of risk management is to remove the risks, hindrances, or obstacles that deter a company from realizing its 'value-added benefits' or meeting its SCRM objectives.

RISK MANAGEMENT FRAMEWORK = ARCHITECTURE

A critical element of any SCRM is the choice of a risk management framework. ISO 31000 is a common and globally recognized risk management framework that is used to architect SCRM.

ISO 31000 is a risk management framework and process that has been adopted as a national risk management standard by more than 60 countries. The ISO 31000 framework is illustrated in Figure 8.

The ISO 31000 framework and process can be used to:

- Establish organizational context.

- Identify supply chain risk stakeholders, end-product manufacturers, and other interested parties.

- Identify supply chain stakeholder risk requirements, needs, and expectations.

- Identify and establish the context for architecting, designing, deploying, and assuring SCRM.

- Evolve as the guideline to evaluate and manage upside (opportunity) risk and downside (consequence) risk.

- Architect, design, deploy, and assure a SCRM system.

- Support ISO 9000:2015 in the design and deployment of Risk Based Thinking (RBT), Risk Based Decision Making (RBDM), and Risk Based Problem Solving (RBPS) in SCRM.

- Form the basis for RBPS and RBDM, which are the bases for architecting, designing, deploying, and assuring a SCRM initiative.

- Establish the basis and foundation for integrating SCRM with ERM.

- Treat and manage supply chain risks.

- Report and document the results and effectiveness of risk treatment and risk management.

- Communicate the effectiveness of the SCRM framework and risk management process to stakeholders, end-product manufacturers, and interested parties.

- Monitor, review, and assure supply chain risks based on organizational risk criteria and risk appetite.

ISO 31000 was not intended to be a standard for management system certification. ISO 31000 is a risk management guideline.

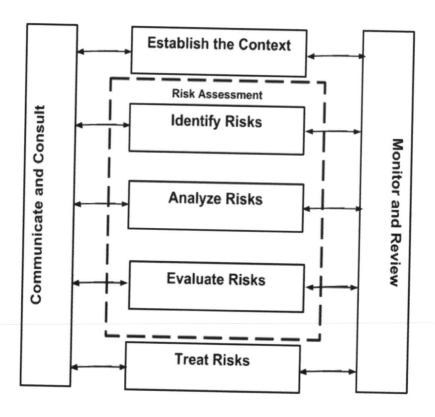

Figure 8: *ISO 31000:2009*

SCRM VALUE PRINCIPLES

A competitive world has two possibilities for you. You can lose. Or, if you want to win, you can change.
Lester Thurow, Management Professor

The market for this book is global. As well, many global end-product manufacturers are using ISO 31000, the international risk management standard as their SCRM framework and architecture. So, many of the discussions in this book are based on this international standard, which is shown in the top figure.

ISO 31000-2009 has eleven risk management principles that can form the basis of a SCRM initiative. The principles are:

   1. SCRM establishes and sustains value.

2. SCRM is an integral part of all organizational processes.

3. SCRM is part of decision making (RBDM).

4. SCRM explicitly addresses uncertainty.

5. SCRM is systematic, structured, and timely.

6. SCRM is based on the best available information.

7. SCRM is tailored.

8. SCRM takes human and cultural factors into account.

9. SCRM is transparent and inclusive.

10. SCRM is dynamic, iterative, and responsive to change.

11. SCRM facilitates continual innovation of the organization.

The successful implementation of these SCRM principles determines the architecture, design, deployment, and assurance of the SCRM process:

1. SCRM ESTABLISHES AND SUSTAINS VALUE

SCRM ensures that value is created by identifying opportunities (upside risk) for investment, mergers, or acquisition. SCRM is critical to the innovation of operational excellence, supply management, product development, and product quality.

SCRM also consists of processes to identify the inhibitors, constraints, and roadblocks (downside risks) that get in the way of meeting quality, cost, schedule, and risk objectives. SCRM also helps ensure success by mitigating catastrophic events that may impede achieving a business objective.

2. SCRM IS AN INTEGRAL PART OF ALL ORGANIZATIONAL PROCESSES

SCRM is an essential part of all organizational processes because it involves RBPS and RBDM. SCRM is also key to all management processes. Most if not all supply chain problem solving (RBPS) and decision making (RBDM) involve uncertainty, which give rise to risk. SCRM creates clarity and sense of purpose to solve problems and achieve objectives under uncertainty.

SCRM can be used in varied and complex settings. SCRM provides analytic tools to assess, analyze, mitigate, treat, and control risks within the organizational context. As well, context provides a lens by which SCRM can be viewed. Context or environment becomes important when evaluating Tone at the Top, management style, and skills to be an effective leader and manager.

3. SCRM IS PART OF DECISION MAKING (RBDM)

RBPS and RBDM are integral to the organization. The purpose of SCRM is to provide decision makers with the right information at the right time to make better choices within the organizational risk appetite. Decision makers use SCRM to make better choices regarding opportunity risk (upside risk) and consequence risk (downside risk).

Decision making (RBDM) is described as a process of selecting the best or optimal course of action among several alternatives, options, or choices. Decision making involves several options. Decision making may involve identifying and selecting supply alternatives based upon the organizational context, culture, and preferences of the decision makers. Decision making may also involve deciding not to do something. Decision making may involve determining how to treat or control risk, specifically whether to accept, share, control, or transfer it (RBDM).

4. SCRM EXPLICITLY ADDRESSES UNCERTAINTY

We discussed Volatility, Uncertainty, Complexity, and Ambiguity (VUCA) in a previous chapter. Uncertainty is also key to RBPS and RBDM. Risk owners should understand the significance, sources, types, and elements of risk.

Uncertainty is a critical element in each step of the SCRM journey. ISO 31000 defines risk as "effect of uncertainty on objectives." This is a broad statement because business and even much of life is governed by uncertainty.

In ISO 31000 and other standards, once business objectives are identified, risks are assessed, controlled, and mitigated. The challenge is much supply chain uncertainty is not well understood and sometimes cannot be well defined.

5. SCRM IS SYSTEMATIC, STRUCTURED, AND TIMELY

SCRM is composed of three RBPS and RBDM attributes, specifically:

- **Systematic.** Systematic means doing something according to a plan or method, such as the SCRM system and risk management framework. The development and application of SCRM also follows a well-defined and thought out method.

- **Structured.** Structured means SCRM RBPS and RBDM follow a replicable and scalable process, such as the Deming Plan-Do-Check-Act (PDCA) cycle or ISO 31000 framework.

- **Timely.** Timely means the SCRM system follows a logical cycle so problems are solved and decisions are made in the right sequence and at the best time. If decisions are made too early or too late then opportunities may be lost or additional costs are incurred (RBDM).

6. SCRM IS BASED ON THE BEST AVAILABLE INFORMATION

Accurate, reliable, sufficient, and suitable information about supply chain context, events, risks, and sources of risk are available for reliable RBPS and RBDM. This is a critical principle because the sources of data may be unreliable – all of which creates uncertainty.

SCRM is a process with inputs, activities, and outcomes. SCRM is dependent on the accuracy, reliability, and consistency of available information. If the quality of SCRM problem solving (RBPS) and decision making (RBDM) information are questionable, then outcomes will be questionable. This is a risk example of garbage in and garbage out.

Evidence based, decision making is common in many disciplines and is a subset of RBPS and RBDM. Evidence based decision making involves:

consultation with experts to evaluate inputs, development of a standard decision making process, evaluation of assumptions, and deep understanding of the problem. The hoped-for result is there will be higher accuracy and reliability of the decision (RBPS/RBDM).

7. SCRM IS TAILORED

The architecture, design, deployment, and assurance of SCRM are tailored to the organization. A risk management standard such as ISO 31000 can be used as a SCRM guideline. However, any risk management standard or guideline needs to be designed, adapted, and crafted to the organization and its context. This principle involves sophisticated and deep elements that we are discussed throughout this book.

For example, ISO 31000 is only one risk management framework and architecture for SCRM. The standard offers a generic outline for SCRM design. This may be its greatest strength and potentially its greatest weakness. Why? Since it is generic means it can be applied in any company, environment, and culture. But, it also implies there is more variability in interpretation and application, which may add uncertainty and potential risks in SCRM design and deployment.

8. SCRM TAKES HUMAN AND CULTURAL FACTORS INTO ACCOUNT

SCRM considers organizational risk, maturity, and stakeholder capabilities to ensure that organizational objectives can be met. More often, SCRM must also consider human and cultural factors.

SCRM has a technical and cultural element. Technical element involves architecting, designing, deploying, and assuring the SCRM framework and process. Cultural and human elements involve behavioral factors that are essential to a successful deployment of the SCRM system.

The human element is often the biggest challenge in SCRM deployment. Why? SCRM involves RBPS and RBDM both of which require people adopting and adapting to new management, controls, and other behavioral changes that people may resist.

9. SCRM IS TRANSPARENT AND INCLUSIVE

SCRM facilitates RBPS and RBDM by everyone in the organization from the Board level to the activity level.   It is critical that stakeholders, end-product manufacturers, and interested parties know how problems are solved and decisions are made. This creates alignment of purpose, mission, vision, and helps ensure objectives are met. It facilitates a common understanding of SCRM objectives and risks that can inhibit meeting these objectives. Transparency also facilitates more effective and efficient use of resources (RBDM).

Inclusiveness is another critical attribute of good SCRM. As part of inclusiveness, an organization can have multiple stakeholders both within and outside the organization. Stakeholders include employees, management, banks, unions, regulatory agencies, suppliers, and other critical parties.

10. SCRM Is Dynamic, Iterative, And Responsive to Change

As external and internal context changes, SCRM must adapt to reflect the changes. This may mean adapting to changes in end-product manufacturer requirements, changes in the business model, changes in the competitive environment, and the uncertainties these entail.

So, let us look at the critical attributes identified in the risk principle:

- **Dynamic.** The VUCA environment, context, and business model for business can change quickly. Organizational objectives may change, which will impact problem solving (RBPS) and decision making (RBDM) processes. SCRM must adapt to new risks, changes in risk appetite, and new risk treatments.

- **Iterative.** SCRM involves iterative processes. For example as more information is gathered regarding a business objective, new risks may be uncovered and new forms of treatment are designed. This is an iterative process so risk-controls and treatment options are deployed in various combinations until risks are within the organization's risk appetite.

- **Responsive to change.** In today's VUCA world, SCRM is responsive to changes. This implies that performance is monitored, reviewed, and revised continuously. SCRM is designed to be flexible. SCRM is also monitored and reviewed to ensure that Key Performance Indicators (KPI's) and Key Risk Indicators (KRI's) are aligned with strategic organizational goals.

11. SCRM FACILITATES CONTINUAL INNOVATION OF THE ORGANIZATION

SCRM has a strategic and tactical perspective that focuses on continual innovation including enhancing competitiveness and ensuring organizational sustainability.

SCRM is designed and deployed so it is used to facilitate continual innovation by parties including government, private companies, for profit organizations, non-for-profit organizations, and individuals.

FINAL-CUSTOMER VALUE ATTRIBUTES

The goal as a company is to have customer service that is not just the best but legendary.
Sam Walton, founder of Walmart

A challenge and priority of end-product manufacturers is to identify final-customer value attributes, develop appropriate customer satisfaction measures, and identify supply chain partners to provide these in an integrated fashion.

CUSTOMER VALUE ATTRIBUTES = SUPPLY CHAIN OBJECTIVES

The end-product manufacturer will determine what product/service value attributes matter to the final-customer and then determine which can or will be made in-house. Once this strategic decision is made, the end-product manufacturer will outsource design, production, delivery, and product service to key supply-partners.

Final-customer requirements are categorized into product value attributes or service value dimensions by the end-product manufacturer, which are communicated to suppliers as objectives or requirements the supplier must meet. Some value attributes are measurable and some are imprecise. For

---

CONTEXT: Airbnb Experience Standards

Airbnb offers vacation rentals and home experiences at people's residences, which are called 'experience hosts.' These hosts according to the Airbnb agreement "enthusiastically share their passion and local knowledge with the world, creating opportunities for guests to gain new knowledge and a different viewpoint." The Airbnb experience standards and attributes are:

- **Credibility.** Hosts are passionate about the experience's theme.
- **Access.** Hosts give guests behind-the-scenes access to people, places, or activities that guests could not find on their own.
- **Perspective.** Experiences tell the host's story so guests become fully immersed in their world.
- **Participation.** Hosts provide opportunities for guests to meaningfully engage in activities or conversations by fully taking part in the experience, not just observing it.
- **Connection.** Hosts are thoughtful about creating environments that feel inclusive and intimate.
- **Accuracy.** Host's description is accurate in terms of the experience in their listing, guests know what to expect.
- **Communication.** Successful hosts are attentive and considerate when communicating with guests, quickly responding when a guest reaches out.
- **Commitment to reservations.** Hosts take care of guest's needs and do not disrupt them.
- **Respect.** Great hosts make each guest feel included and welcomed in the group.
- **Value.** Hosts set a price that balances expectations with what the value of the guest's experience.[65]

---

example, a value attribute, like aesthetics, may not be directly measurable. On the other hand, product performance is probably measurable.

SCRM will then require supply-partners to develop stable and capable design, production, delivery, and service processes. The business goal is to create product consistency, specifically through defining, controlling, and continuously measuring the critical process and product attributes and objectives that matter to the final-customer.

These value attributes and objectives are important because they define what a final-customer wants and define how the value attribute will be delivered to the customer. In the following discussion, you will notice there are two elements: 1. Value attribute and 2. Measure of the value attribute. Both are communicated to critical supply chain stakeholders, especially supply-partners. The ability to consistently deliver these value attributes to final-customers is what drives the value chain.

VALUE ATTRIBUTES

All we are doing is looking at the time line, from the moment the customer gives us an order to the point when we collect the cash. And we are reducing the time line by reducing the non-value adding wastes.
Taiichi Ohno, father of Toyota Production System (TPS)

Value is the real and perceived utility a final-customer or end user has in a product or services. Adding value (upside risk management) and eliminating value detractors (downside risk management) are SCRM objectives.

Value is the real and perceived utility a final-customer or end user has in a product or service. To add final-customer value or utility means that a feature has been added or performance has been improved so the product's image is enhanced and the product is more marketable. SCRM can be thought as designing, deploying, and ultimately managing nine value-related factors:

- Design value.

- Form value.

- Time value.

- Cost value.

- Information value.

- Place value.

- Possession value.

- Perceived value.

- Attribute value.

DESIGN VALUE
Product and service commoditization has altered purchasing patterns. Final-customers more often want similar low online pricing at a brick and mortar retail location. Product, service, and experience design is also at the forefront of retail merchandizing. More retailers are now design-led companies.

One challenge is defining 'what is design'? McKinsey recently asked this question and developed the following response:

> "Like 'strategy' and 'analytics, 'design' is a term that suffers from misuse. Design is not just about making objects pretty. Design is the process of understanding (final) customer needs and then creating a product or service - physical, digital, or both - that addresses their unmet needs. It sounds simple, but it is actually a high bar: the design must simultaneously achieve functional utility, emotional connection, and ease of use, while fitting into customers' broader experience."[66]

FORM VALUE
Form value is the aesthetic, external features of a product. It will be added that a product consists of tangible and intangible elements. Depending on the nature of the product, such as a fashion or consumer product, form is the most important value attribute.

TIME VALUE
Time value means reducing cycle times throughout the supply chain from product development to delivery. More often, time value is associated with having the product available when the customer wants it. This is important to suppliers of manufacturing companies. Firms do not want a large inventory on their books because it is expensive. Today, the goal is to keep all

costs low, including the cost of carrying inventory. End-product manufacturers now ask suppliers to provide products just-in-time to be used or assembled into a finished product.

COST VALUE

Cost value is the overall cost of the product or service over the contract or product lifecycle. Transactional purchasing traditionally dealt with the price of the delivered product at one point in time. Part of the customer-supply partnering assumption is the end-product manufacturer will invest time, resources, and technology with supply-partners who will share cost reductions with the end-product manufacturer.

INFORMATION VALUE

Information value is the ability to have the right information at the right time to make smart decisions. Information value requires the right information technology infrastructure and empowered professionals to make decisions (RBPS and RBDM).

PLACE VALUE

Place value means having right products available where the final-customer wants it or in other words, managing logistics. In the preceding JIT production example, manufacturers want parts delivered directly to the location on the manufacturing line where they will be used. For example, an end-product manufacturer may ask a supplier to deliver parts to loading dock No.10 at 9:00 am on December 6th in Portland, Oregon so they can be assembled into the finished part. Parts delivered earlier or later pose risks to the trucking company and to the end-product manufacturer.

POSSESSION VALUE

Possession value means completing the transaction and gaining possession so the final-customer has the legal right to use the product. Information, warranties, spare parts, and instructions have been provided to the customer so he or she can use the product safely and meet compliance requirements.

PERCEIVED VALUE

Perceived value is the overall perception of a product or service. Often, this involves intangible product and service attributes. Positive perceptions of a supplier may result from its brand, service, quality reputation, supplier reputation, product design, and the many little things that create perceptions.[67]

ATTRIBUTE VALUE

Attribute value consists of the product and service features and functionality that the final-customer wants, requires, and expects. In the next section, we discuss 10 value attributes. These attributes are added, subtracted, substituted, combined, or enhanced to provide specific product, service, or experience value to the final-customer.[68]

EXAMPLES OF SPECIFIC VALUE ATTRIBUTES

Quality is free. It's not a gift, but it's free. The 'unquality' things are what cost money.
Phil Crosby

David Garvin, a Harvard Business School professor, recognized the problem of multiple definitions and identified quality attributes or objectives, which we have extended to the following value dimensions. If these value attributes cannot be achieved to satisfy the final-customer, then a risk arises that must be remediated. If these value attributes are business objectives, then risk-controls are designed so specific value attributes are delivered to the final-customer.

The 10 critical value attributes are:

1. Performance.

2. Reliability.

3. Conformance.

4. Usability.

5. Maintainability.

6. Customer service.

7. Aesthetics.

8. Environmental.

9. Logistics.

10. Software quality.[69]

PERFORMANCE

Performance is the ability of a product to operate up to the expectations of the end user or final-customer. Performance is measured differently in different products. In an automobile, performance includes acceleration, braking, and handling ability. And in a pair of running shoes, it includes branding, comfort, style, durability, and weight.

Performance is a user-based definition of value, which means that an automobile can satisfy different expectations of different users. For example, a racecar driver expects a different level of performance from an Indy Ford compared to a Ford Focus.

Suppliers also are expected to improve a product's performance, packaging, reliability, and maintainability. Innovation may vary based on the type of product, such as whether it is an industrial or consumer product. Let us look at a computer. Making it operate faster and transmit more information may enhance the performance of a personal computer. Packaging is improved by making it more aesthetic (form), easier to operate (user friendly), lightweight, and more compact.

RELIABILITY

Reliability is long-term quality and is the probability of a product failing after a specified period under defined operating conditions. In other words, reliability measures the likelihood that a product or assembly will work over time. Reliability is both a user-based and a conformance-based, value attribute. For example, if a final-customer buys a lamp with a one-year warranty, the customer expects it to last at least one year. Otherwise, the customer will return it for repair, refund, or replacement. The lamp manufacturer knows this and does not want to spend much money replacing defective lamps. Therefore, the lamp is designed and manufactured to last at least one year.

Reliability is measured in terms of mean-time-between-failure (MTBF) or mean-time-to-first failure (MTFF). MTBF is the average time it takes a product to fail between two successive failures. MTFF is the average time it takes a product to fail based on the first time it is put into service. For example, automobile manufacturers know what parts of an automobile will likely fail, so they suggest checking and replacing parts based on a preventive maintenance schedule.

CONFORMANCE

Conformance is the ability of a product to comply with a specification, standard, or design. There are thousands of standards that specify how a product is designed, built, tested, installed, stored, maintained, and repaired. For example, a screw is a simple fastener that holds one part onto another. A screw specification may designate height, width, thread taper, threads per inch, material, and material strength.

Conformance is an engineering or manufacturing definition of product value, because it is related to a measurable product attribute. Conformance is measured in terms of defects rates or levels. For example, a specification for a screw may state that a certain type of screw should have 11 threads per inch and should be made of soft steel. If a screw randomly selected and inspected from a lot has 12 threads per inch and is made of stainless steel, then the screw has two nonconformances. These are product defects or risks, even though the screw has more threads per inch and is made of a stronger material than required by the specification.

USABILITY

Usability is a key product attribute. A product may have all the technical bells and whistles, but will it be user friendly. All of us have been frustrated in trying to use software that simply does not work. The reasons are technical incompatibility, poor documentation, user inability, or a number other reasons. One company that has tested hundreds of software products uncovered many blunders, including excessive technical jargon, overuse of icons, over reliance on training to overcome poor software design, cryptic error messages, and poorly conceived help systems.[70]

MAINTAINABILITY

Maintainability is the same as serviceability. It is the ability of a defective product to be repaired easily, quickly, and economically. For example, if an automotive water pump fails, it is replaced. But first, the automotive equipment in the way of the water pump is removed. Since the compact car was designed to be light weight, the area under the hood is tight and everything is squeezed into a small space. This decreases automobile weight and decreases serviceability because the automobile is harder to work on.

This value attribute is both a conformance and user-based definition. In an engineering sense, maintainability can be measured. A common measurement is mean-time-to-repair (MTTR), which is the average time it takes to repair a defective water pump or service a customer. In a user-based sense, maintainability reflects how the final-customer feels about replacing the defective unit. Was it easy to replace? Did the repair person get his/her knuckles bruised? Could it be done quickly? Or, if a serviceman at a repair station performed the work, was the service courteous, fast, and economical?

CUSTOMER SERVICE

Customer service is an important value attribute used to encourage repeat sales. So, companies design products that can be serviced easily and safely. And companies are providing faster turnaround service to repair defective products. For example, one end-product manufacturer promises it will deliver repair parts anywhere in the world within 48 hours and Mercedes guarantees 24-hour service in California and Arizona.

AESTHETICS

Aesthetics consist of the intangible value attributes: how a product looks, feels, sounds, tastes, or smells. Obviously, aesthetics is subjective, but they are a very important form-based attribute of brand value.

Many people buy a car not based on measurable, technical features of conformance or reliability, but on their value perceptions of quality, fit, and finish of the vehicle. People are often more concerned about the 'fit and finish' of an automobile than the technology inside, such as a new turbocharger. Or, they buy based on current and faddish styling standards. The same can be said for several other major consumer purchases, such as in apparel and fashion sectors.

ENVIRONMENTAL

Environmental groups and regulatory authorities are important stakeholders in the SCRM process. Many perceive that natural resources are used and abused. The world's environment is radically changing as indicated through ozone depletion, unbreathable air in American cities, toxic waste sites, and the list goes on. Sustainability is now the rallying cry of many concerned end-product manufacturers that are listening to green consumers.

Businesses are also listening because of the Valdez oil spill, Three Mile Island, and other environmental disasters. Environmental concerns go well beyond being better corporate citizens, companies want to improve their public-customer perceptions, limit their legal exposure, and manage resources through recycling. Compliance to regulations is moving toward lean management of resources. How is this done? End-product manufacturers are designing highly reliable and maintainable products. Packaging consists of recycled materials. Obsolete products are recycled or disposed cleanly.

LOGISTICS

Logistics is a form of place value - getting the right material to the right location at the right time. However, something this simple is difficult. Something as simple as sending a retail package from one location to another seems to require a logistics analyst and an algorithm to determine the best delivery option.

One company may ship packages using UPS, while another may ship using FedEx or U.S. Postal Service. Each has different package pricing strategies for different countries, confirmation, surcharges, form of transportation, handling charges, insurance, tracking, minimum ground charges, delivery windows, package weight, package dimensions, residential/commercial delivery, and other factors. Frankly, these are confusing and sometimes even seem arbitrary as they change weekly.

SOFTWARE QUALITY

Software is integral to all products. As discussed, we now have 'smart' toasters and washing machines as part of the Internet of Things. Software value, just like any tangible product, must satisfy users' requirements, comply with relevant standards, be cost-effective, and be measurable. For example, software must also be interoperable, intraoperable and exhibit several other value characteristics.

CREATING SCRM VALUE

If you are going to do Toyota Production System, you must do it all the way. You also need to change the way you think. You need to change how you look at things.
Taiichi Ohno, father of the Toyota Production System

Creating and sustaining value becomes one of the most critical elements for developing SCRM innovation and improvement strategies. Value, the opposite of cost or waste, is the only true measure of determining whether a function, activity, or process is providing customer satisfaction, efficiency, effectiveness, or other performance benefits.

MANAGING UPSIDE RISK = VALUE CREATION

In simplest terms, a value chain is the entire design-production-delivery-service process, regardless of which firm owns any particular value-adding step. The value chain encompasses not only first-tier suppliers but suppliers' suppliers, but also end-product manufacturers.[71] The value chain integrates the core processes to provide a seamless chain of value adding opportunities.

Conventional business wisdom now says product quality by itself is not enough to please final-customers. Now, it is the right bundle of price competitive, products, and services delivered just-in-time and in the right manner to the end-product manufacturer and final-customer. In other words, value incorporates form, time, cost, and the other value elements discussed in the prior section. If the end-product manufacturer is not able to provide the requisite value, then risks emerge.

HOW TO ADD SCRM VALUE

More industrial and commercial operations managers as well as consumers are buying based on real and perceived value. According to Mike Treacy and Fred Wiersma in **The Discipline of Market Leaders**, competitive companies identify value-added characteristics, features, or services that final-customers really want and then raise customer expectations in one or two areas of value. These areas of value become the product and service differentiators to the final-customer. Obstacles, hindrances, or risks of achieving or maintaining value are critical supply chain risks that need to be managed and controlled.

Every competitor entering the market is then judged by the market leader's delivery of real value. In other words, a company has three value strategies: 1. Follow the leader; 2. Break out of the pack; or 3. Move ahead of the pack.

HOW TO CREATE TACTICAL VALUE

SCRM often requires a new lens by which to review and evaluate risk factors associated with suppliers. Operations managers still look at suppliers in terms of ISO 9001 compliance, Six Sigma, and quality capabilities. However, many global end-product manufacturers have moved beyond looking at suppliers transactionally or in terms of product specification, or compliance, but are assessing each supplier in terms of its value contribution and risk to the entire value chain.

Once a company has decided on its value strategy, then the innovative company competes on one of three value based disciplines, specifically on:

Operational Value

Operational excellent companies offer middle of the market value at a competitive price with the least convenience. One SCRM truism is to 'drive unnecessary costs and wastes out of the system and return them to the final-customer.' The evidence is all around us. Bandwidth costs are falling. Automobile costs are flat. A product's price/value is reduced through increasing internal efficiencies and lowering operational costs.

In the supply chain, end-product manufacturers and suppliers are collaborating to provide higher value to final-customers. End-product manufacturers are communicating this to all supply chain stakeholders including employees. Employees, even in traditionally adversarial work environments, partner with management to make companies cost competitive. For example, Ford Motor Company asked its employees, United Auto Workers, to find ways to cut auto costs and improve quality. Savings came in small increments but mounted. The goal was to make the car competitive by holding down price increases and offering bigger discounts to final-customers.

Product/Service Leadership

Product leader companies offer products that push cost, performance, and value boundaries. Value can also be added through redesigning a product such as an automobile and collaborating with key suppliers during the development process. Let us look at how vehicle's design value has changed. Fins, chrome, straight lines, curved lines come in and out of design fashion. The jellybean auto dominated most of the 80's and 90's. Performance and sustainability through electrical battery vehicles dominated the 2000's. Now, new automobiles have an edgy, technology look. And within five or so years, autonomous vehicles will enter the market. Also, entry level auto models have technology that was common in three-year older, higher cost vehicles.

Service leadership is equally important. End-product manufacturers want to have a five nines or 99.999% uptime for online services. More often, we see five nines as a service differentiator between data centers or online services. Five nines translate to 5.23 minutes of system downtime a year.

The organization requiring 99.999% up-time requires the data center has processes and systems that are in control, capable of meeting requirements, and improved. The system may have redundancies and operational backup to maintain that level of service. Service life cycle becomes a critical value differentiator for the company. As well, business and IT processes are constantly adjusted, non-value-added work is eliminated, resource utilization is enhanced, work is automated, and risks are eliminated.

Customer Intimacy

Customer intimate companies deliver product and service value that satisfy specific final-customer needs and expectations. Suppliers can add specific product features or personalize a product to enhance product or service value to the final-customer. For example, end-product manufacturers may offer a basic line of automobiles or may offer limited editions with special performance packages to attract upscale consumers. To update a well-

---

CONTEXT: Key SCRM Value Questions

- Does the end-product manufacturer focus on its core competencies and outsource non-critical work to supply-partners?
- Do supply–partners provide appropriate products and services?
- What are the risks and value to each group of supply-partners?
- Are the right suppliers working on the right things?
- Are suppliers induced and compensated in the most appropriate ways?
- Are suppliers provided the appropriate information to manage internal and their supplier risks?
- Is the supply chain organized and managed in the most efficient, effective, and economic manner?

---

known automobile, one manufacturer added features, increased engine performance, changed its name and image, and charged a premium price.

End-product manufacturers and service providers can add value by bundling services and products to provide the final-customer with additional value. For example, Starbucks is the company that got us to spring for a $5 or $6 exotic cup of coffee. Now, it is moving to sandwiches, wine, exotic tea, salads and other 'grab and go' foods. Café Starbucks is popping up in your favorite coffee shop or as a stand-alone restaurant. What is going on? Starbucks wants to turn its caffeinated brand name into a lifestyle brand.

SUMMARY

SCRM is an emerging value risk discipline. Many end-product manufacturers are at the start of the SCRM journey. SCRM is a natural evolution from supply chain management.

A critical element of success is a continual focus on value innovation as the following illustrates:

> "Firms continue to identify core competencies, seeking the inherent benefits - such as improved knowledge depth and organizational learning - of greater focus. Other specific technologies may be essential for competitiveness but it may not be practical to maintain

expertise for them in-house.   The science base may be changing too rapidly, making it risky to be a player in that field. Or, it may be too expensive to maintain technological competence. Or, the product may be too complex for any firm to manage all aspects internally. In each case, outsourcing the technology becomes an alternative, but one that entails risks, especially if suppliers provide critical proprietary capabilities and technology integration skills."[72]

NEXT CHAPTER

In the next chapter, we go a little deeper into Risk Based Thinking (RBT), Risk Based Problem Solving (RBPS), and Risk Based Decision Making (RBDM).

CHAPTER 5:
RISK BASED THINKING AND
DECISION MAKING

WHAT IS THE KEY IDEA IN THIS CHAPTER?
Managing in VUCA time now seems to be a conventional wisdom. The challenge is that there are few rules on what to do and how to manage in the 'new normal.'

The *Wall Street Journal* recently had an article on the increasing decision uncertainty CEO's must face:

> "The unprecedented turbulence is seen in deep political divisions in the U.S. and abroad, uncertainty over the strengthened U.S. dollar, the faster pace of technological change and looming global problems such as mass immigration, terrorism and climate change."[73]

CEO's in the above piece offered the following RBPS and RBDM quotes: "we are making more swift decisions", "we may not get everything right", and "failure happens a lot." Welcome to the new normal for problem solving and decision making (RBDM).

International Organization for Standardization (ISO) has adopted Risk Based Thinking (RBT) into many of its standards. RBT is a key concept in ISO 9001:2015, which is now foundational in SCRM, supply selection, and supply certification.

RISK BASED THINKING

If standards are not formulated systematically at the top, they will be formulated haphazardly and impulsively in the field.
John Biegler

ISO is an acronym for International Organization for Standardization, which is an international standards development organization. ISO 9001:2015 is the world's most commonly adopted standard. ISO 9001 compliance or certification has commonly been used as a baseline certification standard for quality. While ISO 9001:2015 is still a Quality Management System (QMS) standard, it has evolved into a business risk document, which we will be discussing in this chapter.

ISO 9001:2015 AS CRITICAL TO SCRM

End-product manufacturers become registered or certified to ISO 9001:2015. Independent Certification Bodies audit companies based on ISO requirements. If the supplier or company passes the audit, then the company receives an ISO 9001 certification, which is a key element of SCRM.

Too often, we have seen the rationale for ISO 9001:2015 certification based on weak reasoning, such as 'we have been certified for 15 years,' 'we have always done it this way,' 'customers require certification,' and so on. In the ISO 9001:2015, ISO integrated Risk Based Thinking into the standard, which is value attribute for end-product manufacturers.

Depending on SCRM goals, an end-product manufacturer may want to develop 'world-class' SCRM processes. However, many suppliers simply need a basic Quality Management System with minimal risk-controls to prevent deficiencies and to meet end-product manufacturer requirements.

While ISO 9001:2015 is not world-class risk standard, it can form the basis for operational excellence and an entry-level SCRM. It consists of basic quality and risk processes such as Plan-Do-Check-Act, RBT, customer satisfaction, auditing, and design review that are essential elements of SCRM.

```
CONTEXT:  ISO 9001:2015 Table of Contents

1.  Scope.
2.  Normative references.
3.  Terms and conditions
4.  Context of the organization.
5.  Leadership.
6.  Planning.
7.  Support.
8.  Operation.
9.  Performance evaluation.
10. Improvement.
```

RISK BASED THINKING (RBT)

ISO adopted Risk Based Thinking (RBT) in 2015. Why? According to ISO 9001:2015, Risk Based Thinking is:

> "... essential for achieving an effective Quality Management System. ... To conform to the requirements of this International Standard, an organization needs to plan and implement actions to address risks and opportunities. Addressing both risks and opportunities establishes a basis for increasing the effectiveness of the Quality Management System, achieving improved results and preventing negative effects."[74]

Several years ago, end-product manufacturers would mature their quality systems from inspection (Acceptable Quality Level based) to Statistical Process Control (SPC) to quality assurance (QA) to operational excellence. During the quality journey, end-product manufacturers would also adopt lean, Six Sigma, project management, and other operational excellence tools. This process was sometimes referred to as increasing organizational excellence, capability, and maturity.

RBT is part of quality maturity process that has been part of ISO 9001 preventive action. ISO states RBT is already what organizations and people

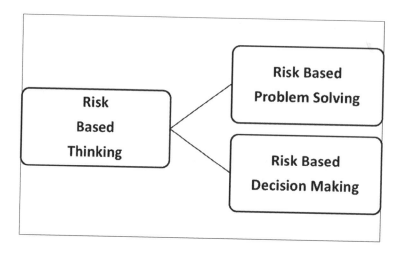

Figure 9: *Risk Based Thinking Components*

automatically do – specifically make risk based decisions. However, ISO 9001:2015 now has explicit RBT requirements that are integrated into a Quality Management System and SCRM.

The SCRM challenge is RBT as defined and described by ISO is difficult to operationalize or audit. How does a person operationalize or audit Risk Based Thinking (RBT)? What evidence, artifacts, or data is the auditor going to find based on someone's thinking? So, how does a person read someone's thoughts? Not unless a person has taken and passed Mind Reading 101 course, one cannot audit or certify Risk Based Thinking.

ISO never defined RBT. In this book, we define RBT as:

1. Risk Based Problem Solving (RBPS).

2. Risk Based Decision Making (RBDM). ™

RISK BASED THINKING IMPACT ON COMPANIES

Many see RBT as a game changer in ISO standards development and SCRM. RBT may impact each family of ISO management system standards. It may also impact operations management professions as well as other ISO standards such as supply management (ISO 28000), information security (ISO 27001), etc.

CONTEXT: Operations Managers Want

- Improved product quality.
- Improved delivery performance.
- Managed inventories.
- Understanding of their business requirements.
- Alternative solutions.
- Help in managing inventories.
- Better communications.
- Customer-supply Risk Based Problem Solving (RBPS).
- Customer-supply Risk Based Decision Making (RBDM)

We believe ISO 9001:2015 is the best thing that has happened to SCRM in 20 years and is a global indicator to the move to SCRM. Why? More than 1.2 million companies globally are registered and certified to ISO 9001 (Quality Management System). As well, more than 400,000 companies are certified to ISO 14001 (Environmental Management System). We estimate that another 400,000 companies are certified to other ISO standards or use the standard for operational improvement. In all, almost 2 million companies are adopting Risk Based Thinking (RBT) and moving to some form of SCRM.

RBT BENEFITS AND USES

ISO 9001:2015 focuses on developing risk based, Quality Management System (QMS) objectives. If the end-product manufacturer can develop QMS financial targets such as profitability, return on investment, return on equity, and risk performance targets, then these will secure executive management support since these are their critical performance metrics/objectives.

As ISO 9001:2015 has evolved into a strategic planning and risk management standard, executive management will be engaged in the QMS and quality reports will achieve higher visibility.

The amount of data is also increasing exponentially due to the Internet of Things and smart devices that are constantly collecting data. Marsh CEO recently said:

"Some 75% or more of the data and analytics companies now use to make decisions was not available just two or three years ago. Such an increase, however, also means there is more complexity in evaluating risks, and in making good choices on strategy and investment."[75]

ISO RBT BENEFITS

ISO standard developers clearly wanted risk to be part of a company's core business mission because of the following benefits:

- Ensure conformance to the ISO 9001:2015 Quality Management System.

- Develop strategic and overarching view of the end-product manufacturer's competitive environment.

- Ensure organizational needs and expectations of interested parties who can impact product specifications and other requirements are met.

- Assure the end-product manufacturer's mission and risk-control approach is based on an analysis of opportunities (upside risks) and well as consequences (downside risks).

- Prioritize risks in strategic planning, specifically providing leadership and commitment planning as well as a review of intended outcomes.

In the next two sections, we will discuss the two critical components of RBT: 1. Risk Based Problem Solving (RBPS) and 2. Risk Based Decision Making (RBDM).

RISK BASED PROBLEM SOLVING (RBPS)

The first piece of intelligent tinkering is to save all the parts.
Paul Ehrlich

In 2015, UPS rebranded from 'We Love Logistics' to 'United Problem Solvers'. While logistics is important, the original tagline conveyed package pickup and delivery. UPS wanted to convey its problem solving abilities:

"UPS 60-second TV spot showcases entrepreneurs, inventors, small-business owners and large enterprises facing a variety of

Risk Based Problem Solving Examples (RBPS)	Develop set of supply management rules
	Calculate 'Make or Buy' Return on Investment
	Identify supplier risks
	Assess supplier risks
	Determine cost/benefits of changing suppliers
	Determine Cost of Quality

Figure 10: ***Risk Based Problem Solving Examples***

challenges, from getting a new business off the ground to expanding operations globally."[76]

In much the same way, we believe that purchasing and supply management must rebrand to Risk Based Problem Solving.

SUPPLY CHAIN SYSTEM

A supply chain consists of interrelated systems and processes such as information systems, risk-controls, logistics systems, financial systems, human resource information systems, planning processes, manufacturing systems, and quality systems. There are also sub-supply systems including internal quality auditing, training, and customer service processes.

Systems theory is the basis for Risk Based Problem Solving (RBPS). Systems problem solving grew out of general systems theory in the last 40 years. Systems theory attempted to define and solve problems by understanding the whole structure (supply chain) and then looking at each component (end-product manufacturer, suppliers, decision points, risks, etc.). Each core internal process or system needs to be understood as well as the interrelationships between the links of the supply chain. Systems theory can form the basis for SCRM and its component parts including RBT, RBPS, and RBDM.

SCRM SYSTEMS APPROACH

The supply chain can be broken down into component parts. The supply chain process consists of many sub-processes and even mini-processes consisting of many steps, each of which has an input and output. An internal supplier provides input to an internal customer who processes and outputs to an internal customer. Each process step should add value in the process chain. Each process step owner is responsible for the value of effort as well as ensuring that downstream customers are satisfied.

Risk management is the method to assure that value is added and sustained at each critical process step. Risk management is also the method to assure that downside risks are minimized through risk-controls at critical steps of the value chain.

RBPS follows a systematic and structured approach focusing on measurable, simple, and doable solutions. RBPS, such as risk assessment, usually incorporates the following elements:

- Identify SCRM innovation project.

- Define problems, risks, or constraints.

- Organize to solve the problem, mitigate risks, or eliminate constraints.

- Identify causes, risks, or constraints.

- Select the optimum solution.

- Deploy the solution.

- Audit solution (RBPS).

Identify SCRM Innovation Project

A supply chain innovation or risk management project is first identified. A project may involve end-product manufacturers or suppliers who have requirements, specifications, needs, expectations, or objectives which are *not* being met. Or, a project may involve improving supply chain effectiveness, adding product functional value, or minimizing process variation.

Safety and health issues usually have the highest priority. Final-customer, market, chronic, major, and cost issues with the greatest potential for gain are then considered. The financial impact of each alternative project is estimated. These are determined by estimating the cost of achieving or not achieving a SCRM objective. Customer surveys, warranty claims, and customer complaints can also identify supply chain innovation projects.

Define Problems, Risks, or Constraints

Supply chain risks and constraints are identified. Risks or constraints are factors that may impede deployment of a solution or meeting a business or SCRM objective. Constraints may include: lack of financial resources, internal resistance, lack of special SCRM skills, or lack of measuring equipment (RBPS).

Organize to Solve Problems, Mitigate Risks, or Eliminate Constraints

SCRM teams are the best approach to solve a problem. A multidisciplinary team has the requisite knowledge, skills, abilities, perspective, and resources to attack, remediate, and risk-control supply problems. Operations management professionals facilitate or lead the team.

Identify Causes, Risks, Or Constraints

Specific supply chain or process owners may be asked to participate in the solution. Those who are impacted by the problem have the best knowledge to identify probable causes. It is important to focus on root-causes, not only on problem symptoms. The root-cause or causes are people, material, methods or machinery or some combination of these.

Select the Optimum Solution

Once root-causes are identified, solutions are proposed to eliminate the root-cause. Brainstorming is one common technique of soliciting solutions from process owners. The 'best' solution is selected for deployment. Best is the optimum solution that is leads to lower costs, removes supply chain constraints, enhances development time-to-market, etc.

Deploy Solution

Deployment is often assigned to those responsible for the problem, the internal or supply process owners. Or, deployment is assigned to the SCRM project team, who performs the work with the blessings of the process owners. Deployment can cross supply, departmental, or functional boundaries (RBPS).

Audit Solution

SCRM innovation is measured to determine the impact of the intervention. Accrued benefits justify the next supply management innovation project. Innovation is then audited to ensure the root-cause or causes are eliminated so the problem does not recur. Once root-causes have been eliminated, a new level of performance, hopefully a breakthrough, has been attained.

Optimize Solution (Start Again)

Once this level has been attained, the supply chain RBPS process starts over again on a new problem. Or, if not enough innovation has been achieved, the process repeats the above steps.

RISK BASED DECISION MAKING

Problem solving and constructive innovation are what business is all about.
Randall Meyer

Managing in VUCA time requires a different lens by which to look at supply chain management and more specifically SCRM. In this section, we discuss the transition from reactive supply management to forward looking SCRM.

REACTIVE MANAGEMENT

Purchasing, quality, engineering, and supply chain management still seem to be mired in reactive management. Reactive management largely involves waiting for an event, hazard, or vulnerability to occur, then the end-product manufacturer reacts to reduce the impact of the hazard; determine and eliminate the symptom; and determine and eliminate root-cause what caused the problem. In many ways, this reflects 'Old School' management

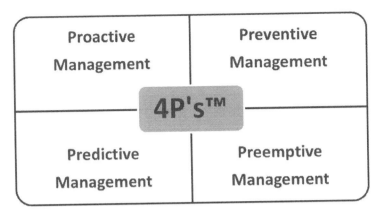

Figure 11: *4P's Management*

that was discussed in Chapter 2. However, VUCA requires 'New School' problem solving (RBPS) and decision making(RBDM).

Ernst & Young (EY) distilled the essence of Risk Based Decision Making (RBDM):

> "Good risk management does not imply avoiding all risks at all cost. It means making informed choices regarding the risks the company wants to take in pursuit of its objectives and the measures to mitigate those risks."[77]

RBDM integrates proactive, predictive, preventive, and even preemptive (4P's™) risk management as shown in the above figure. Proactive, predictive, pre-emptive, and preemptive management are the basis for 'SCRM New School Management.' An end-product manufacturer can respond quickly to unforeseen events and meet critical supply chain objectives. In this way, the end-product manufacturer can focus on identifying and mitigating risks that inhibit its ability to meet a business objective.

CORRECTIVE ACTION MANAGEMENT

Let us first look at corrective action management. Corrective action management involves reacting to a situation and then fixing it. According to the ISO 9000, corrective action is any "action to eliminate the cause of a detected nonconformity or other undesirable situation."[78] Once the root-cause of the nonconformity is discovered, corrective action is undertaken

Risk Based Decison Making Examples (RBDM)	Determine 'Make or Buy' and ROI for each for organization
	Determine appropriate sourcing business model
	Architect, design, deploy, & assure SCRM
	Determine risk management framework to use
	Determine risk appetite and tolerance of organization
	Socialize SCRM with executives and suppliers

Figure 12: *Risk Based Decision Making Examples*

to eliminate the recurrence of the risk, defect, flaw, problem, or discrepancy.

Corrective action may involve:

- Redesigning or modifying the product.

- Redesigning or modifying the process that made the product.

- Training suppliers.

- Updating specifications or changing tolerances.

- Increasing incoming material inspection.

All of the corrective actions listed above are critical, however they still involve looking in the rear-view mirror and fixing something bad or deficient that has already occurred.

PROACTIVE MANAGEMENT

Much of purchasing is still transactional and product based. The end-product manufacturer selects suppliers that they believe can provide conforming products or reliable service. If there is a product nonconformance or poor service, then the end-product manufacturer inspects supplied products or issues a corrective action request. After the fact purchasing and supply management is often acceptable for managing commodity suppliers. However, SCRM techniques are required for specialty products and services.

Proactive management implies anticipatory, change oriented, and self-initiated management and leadership. In other words, proactive behavior involves acting in advance of uncertainty or risk rather than just reacting. It means anticipating risk and taking control to mitigate risks. Taking control means architecting, designing, deploying, and assuring a context sensitive, control environment. It also means making things happen before hand rather than reacting, adjusting, or waiting for a threat or negative supply event to occur.

PREVENTIVE MANAGEMENT

Preventive management is forward looking, specifically looking at the future and seeing what may occur and preventing its occurrence or recurrence. According to ISO 9000, preventive action is any "action to eliminate the cause of a potential nonconformity or other undesirable potential situation."[79]

Preventive management involves designing a risk-control environment, system, and processes to keep undesirable events, threats, or risks from occurring or recurring. Prevention is based on the belief that if a risk is less imminent, then its likelihood and impact are reduced to an acceptable level for the end-product manufacturer.

Corrective and preventive action are sometimes confused. Corrective action is the process of eliminating a nonconformance that has already occurred and ensuring that it does not recur. To do so, the operations manager or process owner must understand the nature of the problem and suggest a solution such as sourcing a new fixture, developing a new procedure, or training employees. Corrective action may result from recurring final-customer complaints, product nonconformances, unstable processes, or product rework (RBPS).

CONTEXT: PPRR Risk Management Model

Queensland, Australia developed the PPRR Risk Model:

- **Prevention.** Take actions to reduce or eliminate the likelihood or effects of an incident by developing a risk management plan.
- **Preparedness.** Take steps before an incident to ensure effective response and recovery to the incident or risk by conducting a business impact analysis.
- **Response.** Contain, control, or minimize the impacts of the incident by preparing an incident response plan.
- **Recovery.** Take steps to minimize the disruption and recovery times by developing a recovery plan.[80]

On the other hand, preventive action attempts to anticipate a potential supply chain problem and identify steps to eliminate its possible occurrence. How can potential supply chain risks be anticipated? One method is to identify high risk areas and ensure there are internal risk-controls to minimize or eliminate the risk (RBPS).

PREDICTIVE MANAGEMENT

Predictive management involves making a statement or judgment about an uncertain or VUCA event. Prediction involves anticipating events or threats before they occur. Systems and processes are in place to ensure business continuity and be able to meet business objectives.

Predictive management involves using past knowledge, big data, trends prior knowledge, or prior experience to anticipate or least understand possible risk futures. A few examples may illustrate prediction. If a supplier consistently has been providing conforming or acceptable quality materials, then supply review is maintained at the same level or even lessened. If the supplier provides a nonconforming or unacceptable lot of supplied material, then product review, inspection, and controls are increased. The common thread to these examples is that history and prior events are precursors to determine the level of acceptable supply risk-control. The challenge is living in a VUCA world makes accurate predictive management more challenging.

PREEMPTIVE MANAGEMENT

Preemptive management involves anticipating and forestalling a threat or risk from occurring by being proactive in the face of VUCA. Preemption is based on the belief that a risk is imminent, material, and consequential to the end-product manufacturer and that some action is taken based on the type and nature of SCRM information available.

The basis for preemptive management is to:

- **Develop business objectives focused on possible/probable outcomes.** End-product manufacturers ensure risks are identified, assessed, and managed using a risk management framework such as ISO 31000, specifically focusing on critical vulnerabilities or obstacles that inhibit meeting critical business objectives.

- **Design a SCRM control environment.** The SCRM control environment is architected, designed, deployed, and assured based on organizational context, which may include products, services, culture, etc. For example, an end-product manufacturer may develop a SCRM control environment similar to what is described in this book. Executive management establishes the risk appetite and tolerance for the end-product manufacturer. SCRM owners are empowered to use RBPS and RBDM to assure critical business objectives are met.

- **Develop an integrated risk methodology and approach across the organization and into the supply chain.** End-product manufacturers can ensure constraints or risks across the supply chain are identified and mitigated so SCRM objectives are met.

FINAL THOUGHTS

Leadership is the art of giving people a platform for spreading ideas that work.
Seth Godin

The American Management Association said this about risk:

> "For virtually every business in the United States, the implications of economic change are enormous. The rapidly changing and more uncertain environment not only has made corporate decision making and planning more difficult, but also has significantly increased business risks. Operating successfully in this new environment will require a very different approach to business management. It involves more, rather than less, attention to external factors, as well as new priorities and strategies and a sharply increased focus on risk management."[81]

Risks are inherent in any customer-supply contract, project, or process. So, smart SCRM is the ability of not being blindsided by these risks or events that can result in these risks. Smart operations managers want to avoid supply crises, firefighting, or reactive management – all of which lead to risk. It is smart RBPS and RBDM to have stable supply processes that focus resources on the areas of highest risk.

The challenge is that RBT, RBPS, and RBDM have become more difficult in VUCA time. As, former Secretary of Defense, Donald Rumsfeld, once said during a press conference, that:

> "There are 'known knowns' – there are things we know we know. We also know that there are 'known unknowns' – that is to say, we know there are some things that we do not know. But there are also 'unknown unknowns' – the ones we don't know we don't know."[82]

IS THERE A BEST FORM OF SCRM MANAGEMENT?

No. We are entering the VUCA age. All end-product manufacturers will respond based on their context, products, final-customers, and business model. What we do know is that more organizations are moving from 'Old School SCRM' to a 'New School SCRM' management designed to their requirements and needs based on RBT, RBPS, and RBDM.

RBPS and RBDM are necessary attributes of good management. One executive at AMN Healthcare was quoted in the *Wall Street Journal* as saying: "We are making more swift decisions and being comfortable with the fact that we may not get everyone exactly right."[83]

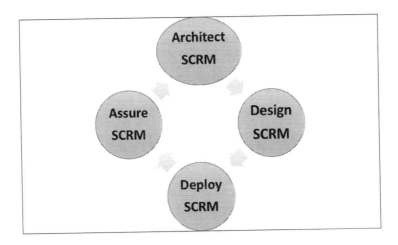

*Figure 13: **Architect-Design-Deploy-Assure SCRM***

End-product manufacturers moving to SCRM incorporate proactive, preventive, predicted, and even preemptive management (4P's) into their RBPS and RBDM. Being proactive means understanding the nature of uncertainty in the types of supplier risks. Being preventive means having contingency plans in place if an unforeseen event materializes. Being predictive means developing processes that are early warning signals of impending threats and risks. Being preemptive means understanding patterns early so hazardous events can be avoided or the possible consequence diminished.

RISK ARCHITECTURE/DESIGN/DEPLOYMENT/ASSURANCE

In this book, we have discussed the importance to architect, design, deploy, and assure a SCRM initiative as shown in the top figure. In the following six chapters, we discuss these topics in more detail, specifically:

- **Chapter 6:** SCRM Architecture and Design #1.

- **Chapter 7:** SCRM Architecture and Design #2.

- **Chapter 8:** SCRM Deployment – Leadership.

- **Chapter 9:** SCRM Deployment – Projectizing.

- **Chapter 10:** SCRM Deployment – Certification.

- **Chapter 11:** SCRM Deployment - Supply Selection.

- **Chapter 12:** SCRM Risk-Assurance.

NEXT CHAPTER

SCRM is a new and evolving discipline. Standards, guidelines, processes, and best practices still need to be developed. In this book, we use 'framework' synonymously with 'architecture.'

SCRM and supply chain risk assessments are usually based on COSO or ISO 31000 risk management frameworks. In the next chapter, we examine how to architect and design SCRM using the ISO 31000 risk management framework.

CHAPTER 6: SCRM ARCHITECTURE AND DESIGN #1

WHAT IS THE KEY IDEA IN THIS CHAPTER

In this chapter, we discuss COSO and ISO 31000 risk management frameworks. They are similar frameworks and architectures. We decided to offer a detailed explanation for SCRM using ISO 31000 since it is based on a global risk management standard. By following each step of the ISO 31000 risk management framework, a company can architect and design a tailored SCRM framework.

SCRM JOURNEY

The only management practice that's now constant is the practice of constantly accommodating to change.
William McGowan, MCI Chairman

How do you start SCRM? There are no fixed rules. The SCRM journey can be product or supplier specific. For example, high-value, high-tech end-product manufacturers would have a different risk management framework than a threaded fastener supplier or distributor.

STARTING THE SCRM JOURNEY

As we discussed a big bang or VUCA disruptor starts the SCRM journey. VUCA disruptors may involve revolutionary and innovative products that may disrupt market assumptions. Big bang disruptors may involve major new products, which offer new styling, have additional convenience, are cheaper, are easier to use, or are faster. Each of these can lead to new challenges and supply risks.

It is critical to first understand and map internal core processes, then move into the supply chain. Internal supply management processes are mapped or flowcharted so that work is understood and disconnects (risks) are identified. This provides problem visibility, provides a sense of process ownership, and encourages cooperation up and down the supply chain. As well, key internal problem solving (RBPS) and decision making (RBDM) points or nodes in the supply chain can be identified.

We recommend SCRM journey starts with a critical few suppliers. As situations arise, corrections are taken to alleviate problems or mitigate risks. Lessons learned become institutionalized and are used to integrate more suppliers into the supply chain.[84]

Risk flowcharting, planning, and other activities are then moved deeper into the supply stream. A Pareto (80-20) analysis focuses efforts on where they have the most impact to minimize risk likelihood and risk consequences. The goal is to communicate mutual wins with suppliers and turn competition into customer-supply coordination and collaboration to work on opportunity risk (upside risk). Making this leap requires a great deal of trust and respect among all parties (RBPS).

Internal and customer-supply process understanding, deployment, and integration are the primary challenges for end-product manufacturers starting the SCRM journey. SCRM is fundamentally process management. The chain is only as strong as the weakest link. So, most SCRM initiatives start with building internal risk capabilities and then integrating these with key suppliers. An end-product manufacturer is then able to say to suppliers: 'do as I do as well as do as I say.'

COSO FRAMEWORK

Eighty-five percent of the reasons for failure are deficiencies in the systems and process rather than the employee. The role of management is to change the process rather than badgering individuals to do better.
W. Edwards Deming

COSO is an acronym for 'Committee of Sponsoring Organizations.'

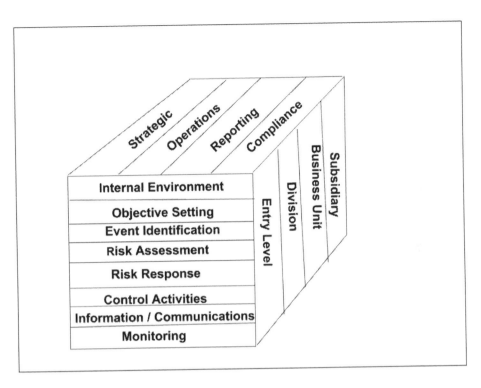

*Figure 14: **COSO ERM Framework***

COSO consists of the following organizations: Institute of Internal Auditors (IIA), Financial Executives International (FEI), Institute of Management Accountants (IMA), American Accounting Association (AAA), and the American Institute of Certified Public Accountants (AICPA). The common element of these associations is they focus on accounting and financial reporting.

COSO's original purpose was to develop a comprehensive framework and provide guidance on how to improve the quality of Internal Control over Financial Reporting (ICFR). The overall goal was to improve organizational performance, risk-control, and oversight. However over the last ten or so years, the framework is now used for operational, IT, cyber security, and SCRM deployment. The COSO ERM framework is shown at the top of this page.

Many companies, usually larger end-product manufacturers, have adopted COSO guidelines for controlling supply chain risks. We call this Internal Control over Operational Reporting. This is a critical point in this book. ISO

31000 and ISO 9001:2015 can also be used as control frameworks. Since these are international standards, the next two chapters focus on how to architect and design ISO 31000 as a risk management framework for SCRM. However, let us first discuss COSO ERM.

COSO ERM

COSO defines ERM as:

> "A process, affected by an entity's Board of Directors, management and other personnel, applied in a strategy setting and across the enterprise, designed to identify potential events that may affect the entity, and manage risk to be within its risk appetite, to provide reasonable assurance regarding the achievement of entity objectives."[85]

This definition of ERM has the following notable components:

- ERM is a process.

- Board of Director's directs, shapes, and oversees the ERM program including organizational risk appetite.

- Executive management owns the ERM program.

- ERM is applied in a strategic setting across the enterprise into the supply chain.

- ERM is designed to identify potential events that can impact the enterprise both positively and negatively.

- Enterprise manages risk within its risk appetite.

- ERM provides reasonable assurance, not absolute assurance, that risks are managed suitably and effectively.

- ERM focuses on achieving business objectives.[86]

COSO RISK MANAGEMENT FRAMEWORK

The COSO ERM cube is designed as a three-dimensional cube or matrix. The cube consists of 8 elements as seen in Figure 14, specifically:

- **Internal environment.** Reflects the end-product manufacturer's ERM and SCRM philosophy, culture, risk appetite, oversight, people development, and ethical values. 'Tone at the Top' is often used as a short cut reference to the state of the internal environment.

- **Objective setting.** Consists of strategic and tactical SCRM objectives, which provide the context for operational risk reporting and compliance. SCRM objectives are aligned with the end-product manufacturer's risk appetite.

- **Event identification.** Identifies potential risks that may positively or negatively impact the end-product manufacturer's ability to design and deploy a SCRM strategy so it can meet its strategic and tactical business objectives. As used in most frameworks, positive risk is considered upside risk or opportunity. Negative risk is considered downside risk or threat or hazard to the enterprise.

- **Risk assessment.** Consists of the SCRM quantitative and qualitative methods, processes, and tools to evaluate the likelihood and consequence of potential events. Common qualitative risk methods include risk maps, turtle diagrams, and FMEA. Common quantitative research methods include SPC charts and statistical analysis.

- **Risk response.** Consists of evaluating various risk response options and their impacts based on risk likelihood and risk consequence. Risk response is the equivalent of risk management. Factors to consider in a risk response may include: cost-benefit analysis; evaluation of variation in project cost, schedule, scope, and quality; and analysis of risk against the end-product manufacturer's risk tolerance.

- **Control activities.** Consists of a system of SCRM policies, procedures, and work instructions that are integrated throughout the end-product manufacturer and into the supply chain. Risk-control activities may consist of process flow charts, decision tools, and project tools dealing with scope, quality, cost, and schedule variances. Risk-control activities ensure business objectives can be met.

- **Information communication.** Consists of the notification and dissemination of critical SCRM information from internal and external sources so responsible parties are aware of the risks and can mitigate them appropriately. Effective communication flows vertically and horizontally in the end-product manufacturer and into the supply chain.

- **Monitoring.** Consists of ongoing activities to provide the appropriate level of risk-assurance and monitoring for the end-product manufacturer. If there is unusual variation within the end-product or service, then the variation is monitored and root-cause corrected so the problem variation does not recur.

The top face of the cube consists of types of objectives specifically strategic, operations, reporting, and compliance. The right face of the cube consists of entity or enterprise components, specifically at the entity, division, business unit, and subsidiary levels.

ISO 31000:2009 RISK MANAGEMENT FRAMEWORK
In the next 2 chapters, we discuss ISO 31000 risk management framework from the perspective of the end-product manufacturer or brand owner. ISO 31000 is shown on the next page in Figure 15.

Specifically, we cover the following ISO 31000 SCRM components:

- Communicate and consult.

- Establish the context.

- Identify risks.

- Analyze risks.

- Evaluate risks.

- Treat risks.

- Monitor and review.

COMMUNICATE AND CONSULT

All we are doing is looking at the time line, from the moment the customer gives us an order to the point when we collect the cash. And we are reducing the time line by reducing the non-value adding wastes.
Taiichi Ohno, father of the Toyota Production System (TPS)

The supply chain starts with identifying final-customer requirements and ends by developing and delivering the right products on time to the final-customer. The final-customer may be an industrial buyer, commercial buyer, or consumer.

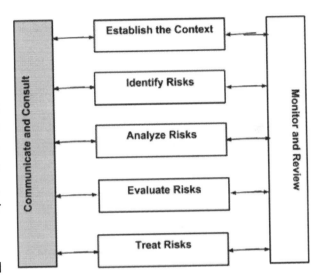

Figure 15: *ISO: 31000 - Communication and Consult*

It is critical that SCRM initiatives, projects, and activities emphasize final-customer requirements and develop strategies on how to satisfy them. Supply management's responsibility is to integrate the end-product manufacturer's core process competencies with the supply base competencies to enhance final-customer satisfaction.

IMPORTANCE OF COMMUNICATIONS

The challenge is too many end-product manufacturers and by inference supply chains are still managed by priorities and politics that do not matter to the final-customer or that generate little income. As well, too many SCRM performance initiatives tend to focus on value attributes that are not important to the final-customer.

Too many companies have discovered that feature-rich, user unfriendly, high-tech gizmo's do not sell. Duh! So, more attention is paid on what features final-customers really want, need, and expect so that optimum revenue is generated. Product designers are finally listening and trying to make technology accessible, understandable, and friendly.[87]

Maintaining a final-customer focus especially for suppliers is difficult. The supplier is far removed from the final-customer. The supplier is providing a component of a part, which is used in a sub-assembly going into an assembly and so on.

Even the best and largest end-product manufacturers sometimes stumble. There are classic examples where companies have tripped. The bigger the company the bigger the stumble appears as the following stories indicate. Coke tinkered with its original formula and depressed sales forced the company to make a public apology as it went back to the original formula. IBM was criticized for its attachment to mainframe computers. General Motors was humbled when it produced full-sized expensive autos when the market clamored for smaller, high-quality, electric vehicles.

SHARED EXPECTATIONS AND INFORMATION

It is important for the end-product manufacturer and its suppliers to have a shared understanding of final-customer expectations. This is developed in a contract, memo of understanding, detailed drawing/specifications, or an expectations roadmap. The memo of understanding outlines the expectations and goals of the supply chain or customer-supply relationship. Specific mutual benchmarks as well as milestones and inducements in the journey are identified for the short, medium, and long-term.

End-product manufacturers are part of a supply chain and have their own supply chains. It is critical that roles, responsibilities, rules, and expectations are developed in the memo of understanding. This will guide SCRM behaviors and shape outcomes.

In advanced SCRM systems, supply chain information monitoring and sharing are automated. The companies throughout the supply chain have identified critical risk-control points and a mechanism for continuously monitoring anomalies, emerging variation, and even early warning signals of impending risk events. As well, extended supply chains are segmented into risk-control areas where risk monitoring and control monitoring are tied to value at risk. The segmenting of the supply chain is based on risk, total supplier spend, product/service criticality, or supplier's history for quality/delivery. The goal is to get an overall understanding and insight into each critical part, link, and control point of the supply chain. A critical control point is a location in the chain where a critical problem is solved (RBPS) or decision is made (RBDM).

A critical element of the supply chain for critical stakeholders is to build trust and share critical SCRM information. Developing mutual trust is difficult. A supplier may not want to share proprietary information with the end-product manufacturer and certainly not with competitors. Most end-product manufacturers have a proprietary business model or core process that they do not want to share with potential competitors. Why? As discussed, a company's core process is what differentiates it from its competition. If all companies had this information, they would lose their value differentiator. So, many operations managers simply say 'no' to full disclosure.

SCRM COMMUNICATION STRATEGIES

Aside from understanding final-customer requirements and communicating these down the supply chain, SCRM communications are used to develop strategies, analyze tactics, and improve operational risk-controls as the following indicate:

- SCRM communications may include sharing information with stakeholders and interested parties on customer requirements; performance objectives; operational metrics/objectives; and project scope of work, and goals.

- SCRM risk maps may be used to fine-tune strategies as well as focus executives and managers on Key Risk Indicators.

- SCRM communications are used to translate SCRM strategy into operational reality using risk dashboards.

- SCRM communications are used to promote Risk Based Problem Solving (RBPS) and Risk Based Decision Making (RBDM) to improve risk-controls, optimize project performance and ultimately set SCRM in the right strategic direction.

- SCRM is used to translate strategy into performance targets and project measures.

- SCRM communications can focus risk mitigation strategies on suppliers that pose unacceptable risks to the end-product manufacturer. Then, operational managers can develop appropriate risk management strategies and mitigation tactics.

- SCRM communications provide SCRM executives and managers a layered view of risk and control information. For example, SCRM is used as an analytical application that allow executives, managers, and users to analyze and explore performance data across multiple levels, specifically, enterprise, programmatic, transactional risks. Risk data users can then access performance baseboard information at any of these levels, from a summarized enterprise to a detailed transactional/product view.

- SCRM communications allow executives to analyze and explore performance across multiple dimensions, specifically technology, process, and people risks. SCRM executives can use the performance data to get at the root-cause of operational problems and issues.

- SCRM communications can also foster communications among organizational stakeholders. Supply executives receive continuous feedback across a range of critical activities enabling executives to steer SCRM.

- SCRM communications are used to analyze the root-cause of critical risks by exploring relevant operational trends by providing SCRM executives with timely information from multiple levels and multiple dimensions at various levels of detail (RBPS).

- SCRM communications can give SCRM executives and managers greater visibility into daily SCRM operations to provide relevant data in a timely fashion and allow for critical demand forecasting.

- SCRM communications can increase coordination among stakeholders to work more closely together. The information can foster a healthy dialog about risk, controls, acceptance of risk, and mitigation strategies among executives and managers.

- SCRM communications can provide a consistent view of SCRM. Enterprise and executive SCRM risk assessment can consolidate and integrate SCRM enterprise and lower level information.

- SCRM communications are used to reduce costs and redundancies through enabling the processing of information and decision making across the agency. SCRM function can consolidate and standardize information and eliminate the need for redundant silos of information, authority, and decision-making (RBDM).

- SCRM communications are developed at the process-owner and transactional level. These empower users by giving them ownership over risks in their area so controls can be designed to mitigate the risks. Process owners can access, analyze the risk information and act on the information.

- SCRM communications can provide actionable information to take actions to mitigate risks, fix problems, and capitalize on new project opportunities before it is too late (RBPS).

SCRM MARKET DRIVEN STRATEGIES

Competitive supply chains are adaptive and harness the energy of bottom-up understanding of market opportunities, technological abilities, team synergies, customer closeness, and top-down leadership, vision, and direction.

There are several excellent strategic supply chain models such as COSO ERM and ISO 31000. The common denominator of most of these risk management frameworks is fanatical customer and market focus.

In **Market Driven Strategies**, George Day outlined the following simple framework of a customer-focused strategy:

- **Arena.** The arena component consists of the markets that an end-product manufacturer and its supply chain wants to target and serve. The arena identifies the final-customers to be served and the critical success factors needed to satisfy these customers.

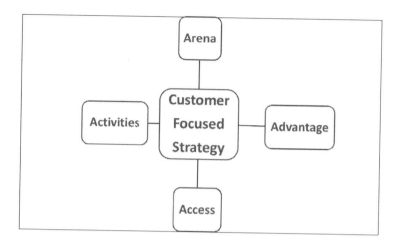

Figure 16: **Market Driven Strategy**

- **Advantage.** Someone recently said that competitive advantage is not between companies but between supply chains. The advantage component consists of the SCRM competencies that differentiate an end-product manufacturer and its supply chain from its competition. The SCRM competencies are also the competitive advantages that an end-product manufacturer must have to attract specific final-customers.

- **Access.** The access component is the SCRM technology, distribution and communication channels that are used to reach final-customers. Marketing channels, specialized niches, supply processes, logistics, and distribution strengths are the means to access specific customers.

- **Activities.** The activities component is what an end-product manufacturer must do to transform SCRM objectives into what a final-customer wants, then identify and eliminate the obstacles, hindrances, and risks that impede meeting the SCRM objectives. [88]

MARKET ORIENTED

Every company is market oriented. What does that mean? Market or final-customer oriented companies deliver cost-competitive, quality products or services. Supply chain processes target specific or mass market segments with defined and coordinated market and final-customer satisfaction strategies. The supply chain challenge is how to induce entrepreneurial customer-supply chains to work cohesively under a unifying final-customer focused strategy.

CUSTOMER INFORMATION HARVESTING

The SCRM process system starts and ends with the final-customer. It starts by identifying final-customer needs, wants, and expectations. It ends by producing a product or delivering a service that satisfies or exceeds these needs, wants, and expectations.

End-product manufacturers are differentiating themselves from the competition by knowing final-customer wants and satisfying these wants. Amazon.com, eBay and other companies are assembling massive amounts of information (also called Big Data) on final-customer preferences based on surveys, buying patterns, etc. The Ritz Carlton developed a detailed database of a customer's past visits so front desk personnel could anticipate customer requirements. Their aim was to know what people buy so they could upsell services directly to hotel visitors based on these preferences.

Final-customer specific purchasing profiles are also common in retail credit card purchasing. Most retail and consumer products are now bar coded and scanned. Let us look at the process. The retail person scans the product. Package labels/codes are tied to a company's sales number in the retailer's system and a receipt is printed for the customer. This point of sale (POS) data relates to the credit card name and is used to develop final-customer profiles. The POS data are then sent to a market databank to analyze, store, and profile individual customer purchasing preferences.

SCRM STRATEGIZING

A vision is a "realistic, credible, attractive future for your organization."[89] The vision is a statement of where the end-product manufacturer, supply chain, plant, department, team or even individual want to be in the future. The vision statement also provides a risk destination the end-product manufacturer and supply chain can aim towards. A vision deals with future, probable outcomes. The vision is energizing. The vision is easy to understand and easy to identify with. The vision will jump-start an end-product manufacturer and supply chain to focus on what needs to be done regarding SCRM.

LINK SCRM WITH THE STRATEGIC VISION

The vision process starts with supply chain stakeholders articulating and then distilling a SCRM direction for the end-product manufacturer and inducing critical supply-partners in this direction. The supply chain vision is then an ongoing, never ending statement that is continually updated depending on marketplace requirements and supply chain opportunities. The transformational leader or CSMO uses the vision to help define where the supply chain is and where it can be in the future.

Examples of common vision statements include the U.S. Constitution, Kennedy's 'we shall put a man on the moon' speech and Ford Motor's End-product manufacturer 'Quality is Job 1.' One end-product manufacturer's vision was 'putting the value in the value chain.' Or, one operations manager recently told me their unwritten supply chain vision is: 'our supply chain can beat yours' (the competitors).

Strategic SCRM planning and RBT form the basis of what the supply chain will look like and what markets it will serve. A well-crafted supply chain vision establishes an attainable benchmark, defines a path, energizes, and encourages, provides meaning to supply owners, is simple and readily understood, and creates a sense of SCRM urgency.

SCRM VISIONING

The SCRM vision requires strong leadership, knowledge of organizational dynamics, sensitivity of supply chain politics, understanding of the competitive marketplace and desire to satisfy final-customer expectations. Once the supply chain vision is developed, operations managers can design and deploy goals, plans, and SCRM objectives. The SCRM vision is then supported through words and deeds by executive management.

Supply chain visioning process will encourage stakeholders to burst out of self-imposed barriers, to think 'outside the supply management box.' Specific issues to consider are:

- **Scope of the SCRM vision.** The SCRM vision is designed to the purpose, competencies, and scope of the SCRM initiative. The vision statement for a corporate supply management function, business unit, plant or SCRM team would be different.

- **SCRM vision of the end-product manufacturer.** The SCRM vision and mission statements will link, echo, and reinforce the organizational strategic ERM vision.

- **Culture of the end-product manufacturer.** The SCRM vision will dovetail with the organizational culture. For example, a SCRM vision emphasizing Darwinist competitiveness is inappropriate in an end-product manufacturer that fosters customer-supply collaboration and partnering.

- **Core risk competencies.** The core risk competencies of the SCRM end-product manufacturer are understood, articulated, and emphasized in the vision and mission statements. Core competencies allow the supply management organization to achieve its vision.

- **Resources.** Sufficient resources are dedicated and available for achieving the SCRM vision.

SCRM MISSION STATEMENT

While similar, a vision and a mission statement are often different. A vision defines the supply chain direction. A SCRM mission defines its purpose, its reason for existing. The mission statement may define, for example, what the supply organization was established to do, such as integrate 'world-class' suppliers into a seamless chain.

The mission statement clarifies SCRM vision, direction, and goals thereby allowing employees and other stakeholders to understand their roles in ensuring success. The mission statement also provides a reference point from which supply SCRM objectives and plans are measured, monitored, and assessed.

The current emphasis is on design-oriented mission statements as the below indicate:

- **Patagonia.** "Build the best product, cause no unnecessary harm, use business to inspire and implement solutions to the environmental crisis."

- **Warby Parker.** "To offer designer eyewear at a revolutionary price, while leading the way for socially-conscious businesses."

- **InvisionApp.** "Question Assumptions. Think Deeply. Iterate as a Lifestyle. Details, Details. Design is Everywhere. Integrity."

ESTABLISH THE CONTEXT

The key to the Toyota Way and what makes Toyota stand out is not any of the individual elements ... But what is important is having all the elements together as a system. It must be practiced every day in a very consistent manner, not in spurts.
Taiichi Ohno

'Establish the context' is the second component of ISO 31000. Organizational context is the environment in which the end-product manufacturer operates, competes, and operates. Context defines the design, deployment, and assurance of the end-product manufacturer's SCRM framework and processes.

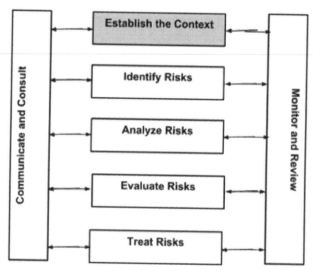

Figure 17: *ISO 31000: Establish Context*

Context is critical to SCRM. Context frames the type, scope, extent, maturity, and capability of the ISO 31000 framework and SCRM journey. Context defines the scope of the SCRM initiative. For example, if the scope in the SCRM program is the entire supply chain including sub-tier suppliers, then the company will find itself trying to boil the ocean.

ESTABLISHING CONTEXT

Context is usually divided into external context and internal context.

External context is the external setting and circumstances in which the enterprise operates and competes. External context describes the environment in which the end-product manufacturer achieves or seeks to achieve its objectives. Examples include: first-tier suppliers, second and lower tier supplier, distributors, logistics, competitors, and collaborators.

Internal context consists of the internal setting, circumstances, and environment in which the end-product manufacturer pursues and achieves its objectives. Internal context consists of operational risks such as plant closure, poor quality products, or lack of inventory.

CULTURE RISK
Some context based risks are internal and external such as culture. Culture is often a strong indicator of future SCRM success. Culture gaps can exist among and within countries. Cultural gaps can exist inside organizations. Culture gaps can exist between individuals. Culture gaps can exist between suppliers.

Culture gaps are often difficult to understand and even perceive. For example, many Westerners still believe that sourcing in Asia is like sourcing in Europe or in United States. Most national economies are interdependent and global. English is often the language of sourcing and business. When we assume that there is a common understanding of business and sourcing protocols. Quite often, this is the furthest from the truth. The supplier 10,000 miles away in a different culture may have different expectations, requirements, and needs than a supplier located 10 miles from the end-product manufacturer. And if a problem arises, the method by which the problem is resolved may be much different than the two parties assumed and expected. For example, the loss of face is a significant element in Asian negotiations (RBPS).

SCRM OBJECTIVES
Business objectives focus on outcomes. Risks are identified, assessed and managed using the risk management framework focusing on the critical vulnerabilities or risks in the supply chain that inhibit meeting critical SCRM objectives.

Once the supply process is stable, meeting its objectives, and improving around performance targets, the operations manager can manage by exception instead of 'fighting fires.'

STRATEGIC PLANNING

A strategic plan focuses on where the end-product manufacturer intends to go. An end-product manufacturer may want to enter new markets, design world-class products, or offer new services. Each activity from conception to delivery has several supply chain components. The job of SCRM is to identify supply components that touch on the overall risk-control strategy and design tactics that mitigate the risks so the business mission, strategic plan, and business and SCRM objectives are achieved. Also, the goal of upside and downside risk management is to align, support, and enhance the SCRM business model.

FROM TACTICAL TO STRATEGIC THINKING

Perhaps the most daunting demand from executive management is designing risk-control strategies. Until very recently, executive management often treated supply/sourcing at the process or even product level. It was the center of much activity, but had little linkage to the enterprise or strategic level of the organization.

In too many end-product manufacturers, supply management is still perceived at a transactional level, specifically 'implementers of buying/sourcing products.' But while executive management appears to be inclined to bestow greater responsibilities to the sourcing function, end-product manufacturers are still not clear what the SCRM authority and responsibility boundaries should look like in their organizations.

STRATEGIC ALIGNMENT

The SCRM professional must architect, design, deploy, and assure critical supply processes, relationships, and contracts with the end-product manufacturer's strategic plan. The operations manager does this by reinforcing organizational values, empowering customer-supply SCRM teams, satisfying internal customers, cutting costs, facilitating inter/intra departmental relations, and improving overall SCRM efficiencies. For example, executive management will never criticize aligned initiatives that support final-customers or improve the end-product manufacturer in cost-effective ways.

A critical approach to support and reinforce supply chain strategic alignment is through developing trust and commitment among operating managers. Operational management support is critical because they have the authority, responsibility, and resources to get SCRM projects started and finished.

SCRM PLANNING

Strategic SCRM plans have a two-year or even longer horizon. They focus on identifying final-customer wants, needs, and expectations regarding quality, delivery, technology, service, and cost variance; reviewing the enterprise 'make/buy' decision; building internal risk capabilities; conducting product/supplier risk analysis; designing contingency plans; reviewing demand schedules; and designing supply risk capabilities (RBDM).

James Morgan, a longtime observer of supply management changes, made the case for SCRM planning:

> "To achieve competitive advantages through integrated supply chain management, businesses need to continue to emphasize their outsourcing strategies. Outsourcing can significantly benefit a firm if it has performed in the context of a strategic plan. Outsourcing adds little competitive benefit, however, to a firm that does not have a clear vision of its core competencies and knowledge of where it can compete versus what it should outsource."[90]

ISO 31000 is used in this book to architect and design SCRM. Other risk management frameworks can also be used. If the end-product manufacturer designs and deploys each of the 7 steps in the ISO 31000 framework, the company can develop the beginnings of a SCRM strategic plan, then develop sub-plans.

PURPOSE OF SCRM PLANNING

Supply chain planning is the process of selecting SCRM objectives and establishing a road map for meeting SCRM objectives. Supply chain planning involves deciding where the SCRM initiative should/can go, how to get there, who will lead it, what vehicles will be used, what are the milestones in the journey, and what are the expected SCRM objectives upon arrival.

The SCRM road map is a guideline document or a formal project document identifying final-customer requirements; prioritizing stakeholder requirements; developing a supplier list; defining cost/delivery/quality requirements; detailing supporting documentation; developing robust designs, bills of material, specifications; and so on.

Many end-product manufacturers still focus on product, tactical, or transactional supply planning emphasizing commodity forecasting, cost analysis, quality inspection, or product testing. While these are important, the SCRM should deal with customer satisfaction, supplier capabilities, make/buy decisions, supply development, and information systems sometimes may not be addressed. Strategic SCRM objectives that get management's immediate attention include: revenue enhancement, cost reduction, innovative product development, and most importantly final-customer satisfaction (RBDM).

WHY IS STRATEGIC PLANNING SO IMPORTANT?

The *Economist Magazine* said:

> "... companies are stumbling to find new ways to manage their relationships with their customers. ... Only happy customers will be loyal ones – and loyalty is something companies desperately need if they are to survive in today's difficult economic climate."[91]

SCRM project planning is critical because it:

- **Provides direction.** SCRM planning provides direction for the end-product manufacturer to deploy the initiative. Once an overall strategic direction is established, then more detailed tactical supply plans can be formulated for divisions, departments, teams, and product lines.

- **Provides a structured framework.** SCRM planning provides a structured framework for SCRM decisions (RBDM). Without a plan, the supply chain risk initiative is rudderless and cannot achieve its goals. Each business unit or operational department may interpret a supply chain requirement differently, which affects deployment. A structured framework unifies different risk interpretations, goals,

CONTEXT: KPMG Top 10 Supply Management Risks

1. Ignoring the financial health of private suppliers.
2. Overlooking shifting political tides.
3. Neglecting to monitor early warnings for quality, consistency, and reliability.
4. Underestimating supplier capacity constraints.
5. Assuming all suppliers move at your clock speed.
6. Using more brute force and less collaboration with suppliers.
7. Assuming cyber security is an IT problem.
8. Reinventing the wheel by building and managing an in-house vendor intelligence network.
9. Pursuing an isolationist strategy.
10. Living in the dark.[93]

and tactics into a common SCRM effort that focuses on final-customer satisfaction.

- **Reveals opportunities.** SCRM planning reveals opportunities to improve supply chain efficiencies, reduce costs, improve profits, increase market share, or please final-customers.

- **Facilitates supply risk-control and risk-assurance.** SCRM contingency supply chain planning may anticipate potential problems. Once areas of potential problems are uncovered, plans, risk-controls, risk mitigations, corrective actions, preventive actions, and recovery plans are developed to prevent their recurrence (RBPS).[92]

TACTICAL PLANNING

Once strategic SCRM plans are established, then SCRM tactical plans can be designed. Tactical plans detail how the SCRM strategy will be deployed and assured. Tactics are often systematic plans for action.

Tactical SCRM plans are detailed, specific, and short-term. Tactical plans may detail how a product is made or a service is delivered, detail how a product or service is sourced, detail how SCRM objectives are met; and explain how risk-controls are designed to meet objectives. In general, tactical SCRM plans can specify what business areas or even product lines

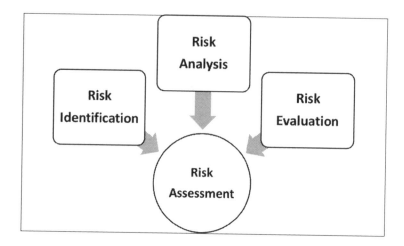

Figure 18: **Risk Assessment Elements**

are developed and how suppliers will be selected, developed, and improved.

KEY PLANNING ELEMENTS

The following are key elements in all SCRM planning:

- **Flexible.** Any journey has switchbacks and dead ends. A SCRM plan is flexible so changes can be made easily.

- **Doable.** Plan outlines a workable and attainable route for the manufacturer as well as key supply-partners for SCRM design, deployment, and assurance.

- **Realistic.** Plan is realistic given the internal and sourcing capabilities and resources.

- **Team based.** Plan incorporates key stakeholders throughout the supply chain and throughout the product or contract lifecycle.

RISK ASSESSMENT

Many of our best opportunities were created out of necessity.
Sam Walton, founder of Walmart

The Institute of Internal Auditors (the IIA) defines risk assessment as:

"Risk assessment is a systematic process for assessing and integrating professional judgments about probably adverse conditions and/or events. The risk assessment process should provide a means of organizing and integrating professional judgments for development of the organization's work schedule."[94]

ISO 31000 RISK ASSESSMENT ELEMENTS

Risk assessment involves three discrete components according to ISO 31000. These elements are covered in the next 3 sections:

- **Risk identification.** Involves identifying potential events that can positively impact the enterprise (upside opportunity) or negatively impact (downside risk) the enterprise's ability to achieve its business and SCRM objectives. An end-product manufacturer identifies the sources of risk, areas of impacts, events, their causes, and potential consequences based on the achievement of business and SCRM objectives.

- **Risk Analysis.** Involves determining how risks are treated or mitigated based on the likelihood and consequence of specific risks occurring. An end-product manufacturer analyzes risk in terms of considering the causes and sources of risk, positive/negative consequences, and likelihood/probability of occurrence.

- **Risk Evaluation.** Involves making the appropriate risk decisions (RBDM) based on context and the risk appetite of the end-product manufacturer. An end-product manufacturer evaluates risk to assist in solving problems (RBPS) and making better decisions (RBDM) at the 1. Enterprise level; 2. Programmatic/Project/Process level; or 3. Transactional/Product level.

SUPPLY RISK ASSESSMENT

Supply risk assessment involves the evaluation of events that may disrupt the supply chain. Internal context as discussed in the previous section may include variability or risk in the delivery, cost, scope, or scope of projects.

Supply chains by their very nature are complex involving many companies, distribution centers, warehouses, production facilities and stakeholders.

Simplification, a form of risk mitigation, is a risk based approach to eliminate programmatic, project, process, and product complexity. Complexity as part of VUCA is a risk. More specifically, these complexities are risks that may result in supply chain disruption, increase delivery schedule, or increase time-to-market product development.

An important ability in assessing SCRM is detecting variation or deviation in supply management plans. The deviations may involve quality, scope, schedule, cost, technology, specification, and other types of critical deviation. This is one reason we are seeing more suppliers managed as a project; critical work breakdown structure involving cost, schedule, and quality; and specific SCRM performance objectives. Any variation or departure from a plan, cost, schedule, or quality target is considered a risk that may have to be controlled and mitigated. Interestingly over the last 5 or so years, food, pharmaceuticals, electric power, and many sectors have developed their own risk assessment requirements and specifications.

Over the next five years, machine learning (subset of artificial intelligence) will have more impact on SCRM. SCRM machine learning will allow operation managers to obtain real-time supplier information to allow for more accurate problem solving (RBPS) and decision making (RBDM).

RISK ASSESSMENT TECHNIQUES

The supply chain risk classification can be qualitative or quantitative risk assessment. A quantitative assessment is based upon statistically accurate history of supplier performance. A qualitative assessment is based upon perceptions of likelihood and consequence in terms of not meeting a supply chain Key Risk Indicator or key performance indicator.

There are many risk assessment techniques. The ISO 31010 standard lists more than 30 techniques as shown in the table below.

LIST OF RISK ASSESSMENT TECHNIQUES		
Method	**Description**	**Application**
Checklist	Simple and quick identification of possible risk uncertainties.	Used in varied ways. Checklist assess-

		ments. Low complexity. Designed to specific application.
Preliminary hazard analysis	Objective is to identify hazardous situations.	Used for threat analysis and cyber security, etc.
Structured interview and brainstorming	Objective is to collect ideas, rank, and evaluate them.	Used for risk auditing
Delphi method	System for combining expert opinions about probability and likelihood in the risk assessment.	Used for collaborative risk assessments.
Structured 'what if'	System approach used by a team to identify and assess risks.	Used in facilitated workshop.
Human reliability	Objective is to understand ergonomic and human system performance.	Used to understand human reliability and risks.
Root-cause analysis	Objective is to understand root-cause of a single loss.	Used in single loss analysis. Medium complexity
Scenario analysis	Identifies future scenarios through extrapolation of the present.	Used to envision future risks. Qualitative approach.
Toxicological risk assessment	Hazards are identified and analyzed including pathways.	Used to comply with regulatory requirements. Specific application.
Business impact analysis	Analysis of key disruption risks that can impact business continuity.	Used in critical applications.

Fault tree analysis	High risk events are identified and lower level risks prioritized. Mitigations are assigned to risks.	Used in many risk applications.
Event tree analysis	Inductive reasoning to translate event likelihood into possible outcomes.	Used with previous tools in multiple applications.
Cause/ consequence analysis	Combination of fault tree and event tree analysis.	Used in multiple applications from first/second/third-party assessments.
Cause/effect analysis	Effect can have number of causes that are analyzed.	Often used with other assessment techniques.
Failure Mode and Effects Analysis (FMEA)	Analysis of failure modes and effects, which are then mitigated/treated.	Used mainly at the product level to ID possible design failures.
Reliability centered maintenance	Method to analyze maintainability failures, safety, availability, and operational economy.	Used mainly for operational risk assessments.
Sneak analysis	Method to identify design problems. Sneak condition refers to a latent hardware or software unwanted event.	Used mainly in product design.
Hazard and operability studies (HAZOP)	Process of risk identification of possible deviation of intended operation.	Used in operational analysis.
Hazard analysis and critical control points (HACCP)	Process to assure product quality, reliability, and safety of processes.	Used in food safety and similar areas.

Layers of protection analysis (LOPA)	Process to analyze control effectiveness.	Used in operational control effectiveness analysis.
Bow tie analysis	Visual qualitative analysis of pathways and causes of risks.	Used in product and process levels. Multiple uses.
Markov analysis	Quantitative analysis of complex systems	Used in repairable electronic and mechanical systems.
Monte Carlo analysis	Process to analyze variations in systems	Used in complex systems.
Bayesian analysis	Quantitative statistical analysis of distribution of data.	Used where sufficient data is known.

IDENTIFY RISKS

There is no such thing as 'zero risk.'
William Driver

The next step is to 'identify risks.' Two general approaches are used to identify risks: 1. Event based and 2. Business objective based.

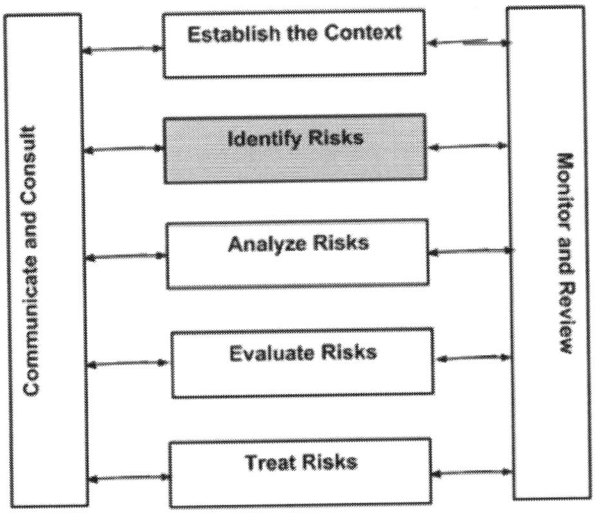

In an event based approach, the operations manager identifies and prioritizes risks of an unwanted event occurring. The SCRM team then develops a plan to

Figure 19: *ISO 31000: Identify Risks*

monitor, prevent, and if necessary mitigate the consequences of the supply chain risk event. Examples of events are hurricanes, earthquakes, supplier labor turmoil, and plant shutdowns. In general, risks are based on a single event or Black Swan occurring, such as a hurricane or earthquake.

In the business objective approach, the end-product manufacturer's SCRM team identifies critical final-customer needs and operationalizes these requirements as SCRM objectives The SCRM team then identifies and mitigates supply constraints, variances, and weaknesses that may inhibit satisfying these requirements or meeting the SCRM objectives. Examples of SCRM objectives include supplier performance, quality metrics, cost targets, or delivery schedules.

The risk register of an end-product manufacturer identifies risks for the overall supply chain and the risks of specific suppliers. What type of risks are identified? The risks identified are those that inhibit the achievement of specific SCRM objectives or are event based risks.

RISK IDENTIFICATION PROCESS

At a simple level, supply chain risk is understood by identifying possible sources of unwanted variation from defined supplier targets and SCRM objectives. In other words, the supply chain is a system composed of processes and sub-processes. Possible risks are identified and controls are designed and deployed to mitigate these risks.

The following are introductory steps for identifying supply chain risks:

- Flowchart the overall supply chain and sub-processes (RBPS).

- Identify possible events that may disrupt the supply chain.

- Define critical supply chain system, process, and product activities that may inhibit meeting SCRM objectives (RBPS).

- Identify probable and possible sources of unwanted variation.

- Identify which sources of variation represent higher levels of risk through a risk assessment (RBPS).

- Develop system, process, and product control points (to prevent product nonconformances (RBDM).

CONTEXT: Top Supply Chain Risks

- **#1 – Product quality risks.** It is difficult to recover from a shipment of critical nonconforming products. Often these products are scrapped and cannot be reworked since the supplier is 8 or 10 time zones away.

- **#2 - Increased inventory.** Increased incoming, in-process, or final inventory has become a risk because of lean, just-in-time, and no-inventory push. However, additional inventory is often a logical risk mitigation, which can result in additional risks because inventory located at different stages of the supply chain is difficult to manage and is costly to maintain because working capital is tied up.

- **#3 – Black Swan event.** Black Swan event is the third highest perceived supply chain risk. Even though Black Swan event is very low likelihood, it can have very high consequence on product delivery and the viability of the supply chain.[95]

TYPES OF RISK

There are many sourcing risks. The list below is only a partial list of risks to the customer, end-product manufacturer, brand owner, stakeholders, or interested parties:

- **Brand risk.** Risk to the brand if an event occurs with a supplier or distributor that can result in brand dilution or loss of brand value.[96]

- **Business risk.** Caused by factors such as a supplier's financial or management stability or purchase or sale of the supplier. These may result from changes in key personnel, new management, new reporting structures, or reengineered business processes.

- **Changing offshoring economic risk.** Low-cost, offshore suppliers are now medium cost suppliers due to inflation, rising supplier labor costs, or for other reasons. In general, Return on Investment (ROI) from offshoring has diminished from 40% to 15%.

- **Climate change.** Food and electronics from suppliers are impacted by heat.

- **Conflicts of interest risk.** Conflicts of interest may involve kick-backs, bribes, fraud, and other irregularities.

- **Corruption risk.** Business customs are often different around the world. In one country, a financial inducement is considered another country's bribe.

- **Cultural differences.** Cultural differences and understandings are barriers to offshoring. Language, culture, business rules, expectations, and understanding are different from country to country which may increase overall risks. For example in Japan, there is cultural tendency to hide or delay communicating negative business information.

- **Cyber risk.** Software is part of almost every product these days. Software in these products may be corrupted or even may be hackable. Hackers may also intercept critical or Intellectual Property (IP) information

- **Data/tech transfer risk.** Electronic transfer of information dealing with suppliers can reveal IP, trade secrets, and personal information.[97]

- **Demand risk.** Suppliers are buffeted by unpredictable final-customer and end-product manufacturer demands. These may include new requirements or sudden increase of supplied goods or services leading to risks of not fulfilling spikes in end-product manufacturer demand for products.

- **End-product manufacturer risk.** If the end-product manufacturer does not want a product or service, there is a risk of a supplier going out of business.

- **Enterprise risk.** Enterprise risks take many forms such as systemic, dependency, interdependency chronic, material, network, discontinuities, gaps, conflicts of interest, reputational, compliance, and regulatory risks.

- **Environmental risk.** These risks are usually outside the supply chain and are related to economic, social, governmental, and climate factors, including the threat of terrorism. Environmental risk may include: hurricanes, tornadoes, or other extreme weather events that can have a devastating impact on the supply chain.

- **Geopolitical risk.** These may involve a revolution in a supplier located country, price of oil dropping dramatically, or a terrorist event disrupting the supply chain. All of these are unexpected, unknown events that can disrupt the supply chain. Other geopolitical examples may include: insurrection; price of labor; monetary policy; political unrest; currency fluctuations; export restrictions; corruption; border delays; cyber requirements; regulations; domestic investment requirements; legal; security; regulatory; environmental compliance.

- **Hidden cost risk.** The end-product manufacturer requests a firm, fixed, and final bid for the work or product. The supplier or contractor supplies a bid that is accepted by both parties. However, when the work is completed or the products are delivered, there are unforeseen extra expenses.

- **Intellectual Property (IP) risk.** Suppliers of complex products that utilize the end-product manufacturer's IP may steal the IP or reverse engineer products. Intellectual Property is a tempting target of cyber theft because research data, product design, proprietary knowledge, and trade secrets can comprise up to 80% of a company's value.[98]

- **Labor unrest risk.** Labor in developing countries is fighting for fairness and equity. Labor unrest can shut a plant resulting in the inability to meet end-product manufacturer demand for products.

- **Lack of supplier visibility risk.** Risk of not knowing what second, third, and fourth-tier suppliers are doing.

- **Language risk.** While English is often the language of offshoring and supply management, language barriers can still exist with critical products.

- **Logistic risk.** High-value cargo, such as electronics, smart phones, and big screen TVs are popular theft items. Ships, trains, and trucks are still hijacked by pirates. Containers may not be secure resulting in stolen products. Or, if there is an event such as a tsunami in Thailand or earthquake in Japan, the transportation of goods may need to be rerouted.

- **Natural disaster risk.** Hurricanes and storms can disrupt supply chains by slowing or disrupting logistics.

- **Management turnover risk.** End-product manufacturer and supply management turnover is a risk since new management may issue additional requirements or specifications that the supplier may not be able to meet.

- **Manufacturing risk.** These are often quality related. The supplier changes production material, methods, or people that results in nonconforming products being sent to the end-product manufacturer.

- **Natural disaster risk.** Natural disaster risks include: climate change; reliance on oil; earthquakes; tsunamis; and storms.

- **Operational capability risk.** Operational capability risks include unexpected changes in production equipment, people, supervision, plant, property, and equipment.

- **People risk.** People risks include: labor disputes; training; corruption; trade secrets; personal data; safety; fraud; kickbacks; and conflicts of interest.

- **Perception risk.** Most end-product manufacturers have been operating continuously and successfully for many years. Then, a Black Swan occurs such as a cyber security breach or fraud. This reflects poorly on management's risk-controls. The investment community reappraises and revalues the company lower.

- **Physical plant risk.** Supply chain may break down because of the condition of a supplier's physical facility due to a lack of maintenance or something similar.[99]

- **Planning and control risk**. These are caused by inadequate supply and demand planning, which amount to ineffective SCRM.

- **Poor supplier quality risk**. Supplier failure to deliver quality and conforming products is often the #1 supply chain risk.

- **Process risk.** Supplier manufacturing processes may not be stable or capable of making products to specifications. Design processes may not be capable of developing final-customer pleasing, reliable products.

- **Production risk.** This is similar to supplier quality and reliability risk. The supplier may not have the capability, risk-controls, or maturity to meet the end-product manufacturer's quality requirements.

- **Product/Service risk.** Product/service risk may include: poor product quality; lean delivery; short product shelf life; untenable service level agreements; excessive handling; variances in delivery; variances in costs; lack of product/service value; short/long product lifecycles; transportation risks; difficult export rules; technology requirements; malicious software; counterfeit parts; hacking; or product shortages.

- **Project risk.** SCRM projects are managed to mitigate scope, quality, schedule, and cost variance risks. Variances outside the required target represent risk to the end-product manufacturer.

- **Reputational risk.** Human rights violations, environmental degradation, health and safety issues, and poor supplier working conditions can diminish global brand value.

- **Return on Investment (ROI) risk.** Each supplier represents an investment of time, effort, resources, and capital. If there are variances in supplier capability, the end-product manufacturer must invest to mitigate supplier risks, which lowers ROI.

- **Scale supply chain risk.** If the end-product manufacturer wants increased capacity from the supplier, then this can result in longer time frames to replicate plant controls to make similar products or be able to find new suppliers.

- **SCRM risk.** These risks entail not managing domestic or offshore suppliers effectively. These are caused by not putting contingencies (or alternative solutions) in place if something goes wrong.

- **Second or higher tier supplier risk.** Second or higher tier suppliers may not be complying with social equity laws such as using child labor which may impact the brand owner or the end-product manufacturer.

- **Sourcing model risk.** Risks of changing between outsourcing vs. insourcing (manufacturing internally).

- **Supplier financial risk.** Financial condition of foreign suppliers can be difficult to ascertain. Accounting and financial statements may not be equivalent across countries.

- **Supply selection/development risk.** Supply selection takes time. Supply development to meet end-product manufacturer requirements can take even longer. Both are risks when products need to be designed, manufactured, and shipped just in time.

- **Supply risk.** This is a broad category that may entail any interruption to the flow of products, whether raw material or parts, within the supply chain. This supplier category may include: financial challenges; lack of mutual trust; poor communications; no buffer inventories; little collaboration; multiple tiers of suppliers; sub-tier suppliers; poor monitoring; no risk-assurance; lack of requisite technology; dependency on a single-source; responsiveness of suppliers; bull whip effects; demand shocks; inflexibility; lack of business investment; lack of sub-tier monitoring and knowledge; single-sourcing; low supplier maturity and capability; high capacity utilization; holding costs; supply shifting risks; poor forecasting; and short/long-terms contracting.

- **Technology risk.** Technology advances are among the most cited global trends that will transform or disrupt the supply chain. Technology risk may include: cloud security; data transfer; lack of cyber security; communication disruptions; lack of shared information; dissimilar IT systems; low capability and maturity; and low investment in IT.

- **Terrorism risk.** Terrorism can result in critical supply chain insta-bility, which can disrupt the supply chain. According to reports, civil unrest is a major supply risk in more than 30% of countries.[100]

ANALYZE RISKS

The graveyard of business is littered with companies that failed to recognize the inevitable changes.
Anonymous

The next step is to 'ana-lyze risks.' Analyzing supply chain risks in-volves reviewing and even investigating sup-plied products but also looking at supply pro-cesses, project controls, programmatic controls and eventually enter-prise controls. As a company moves up the maturity curve, end-product manufacturers may look to the supplier

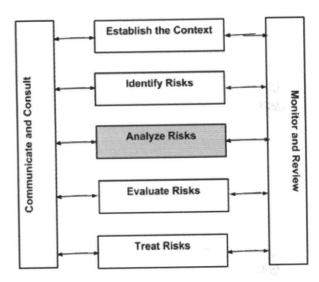

Figure 20: *ISO 31000: Analyze Risks*

to establish and maintain a SCRM or even an ERM program.

Risk assessment can be a qualitative or quantitative method to evaluate risk in terms of the likelihood and consequence of a potential event occur-ring. The RBT and risk assessment tools covered in this chapter are mainly qualitative.

Why do we spend time presenting qualitative tools? Most end-product manufacturers transitioning to or starting their SCRM journey will use a simple risk assessment methodology. And often, there is no one right way to start.

WHAT IMPACTS RISK?

To a large extent, risk is based on perceptions of what may occur and the impact (consequence) of that occurrence to the end-product manufacturer, stakeholders, and suppliers. The operations or supply manager needs the knowledge to understand supply chain processes and assess risks. Several factors can affect the operations manager's perception of risk, including:

- Supply chain technology, design, and manufacturing processes.

- Criticality of supplied products or services.

- Knowledge of supplier KPI's and KRI's.

- Amount of software in supplied products.

- Supplier risk-controls at critical decision points.

- Supplier risk maturity and capability.

- History of early/late deliveries and other performance indicators.

- History of cost overruns.

- Process management, risk-assurance, and risk-controls.

- Adequate documentation including policies, procedures, and instructions.

- Product nonconformances or rejected shipments.

- Amount of risk information available from suppliers (RBDM).

SUMMARY

COSO and ISO 31000 are risk management frameworks. End-product manufacturers starting their SCRM journey may adopt, adapt, and deploy either of these risk management frameworks. Also, companies will develop tailored risk taxonomies to fit their specific context.

They can then architect, design, deploy, and assure their risk management systems based on their context. Also much like quality or lean is a journey, risk management is a journey that starts with end-product manufacturers adopting RBT, RBPS, and RBDM moving towards increased risk maturity and capability.

NEXT CHAPTER

In the next chapter, we continue to examine how to architect and design SCRM using the following components of the ISO 31000 risk management framework:

- Evaluate risks.
- Treat risks.
- Monitor and review.

CHAPTER 7:

SCRM ARCHITECTURE AND

DESIGN #2

WHAT IS THE KEY IDEA IN THIS CHAPTER?

In this chapter, we continue to use ISO 31000 as the basis for architecting and designing a SCRM process.

EVALUATE RISKS

Tell me and I will forget, show me and I may remember, involve me and I'll understand.
Chinese Proverb

TWO ELEMENTS OF RISK QUANTIFICATION

There are two elements of supply chain risk: 1. Likelihood or probability of an unwanted occurrence and 2. Impact or consequence of the occurrence or exposure. For example, there is a 5% probability or chance of a relatively minor occurrence. This occurrence has little impact on the process, project, or product. On the other hand, a 1% probability of a major

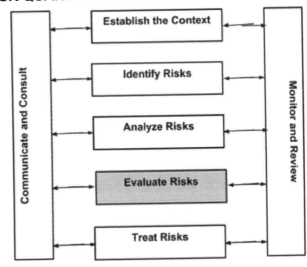

Figure 21: *ISO 31000 - Evaluate Risks*

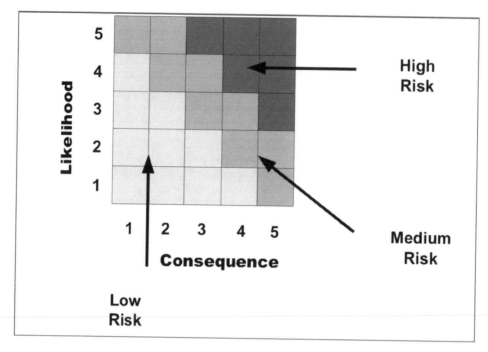

*Figure 22 - **Heat Map - Risk Map***

occurrence such as the inability to meet a delivery window would close down the end-product manufacturer. This is clearly unacceptable.

HEAT MAPS

Many companies start the SCRM journey use qualitative tools such as risk or heat maps as the basis for risk analysis. The risk map methodology is commonly used in many organizations, however it must be designed and tailored to the end-product manufacturer's context to increase its effectiveness and efficiency.

A heat map is a visual representation of risk as shown in the above Figure 22. Each risk is scored and prioritized on a 5 X 5 Risk Exposure Matrix ranking *likelihood (probability)* of the risk occurring against the *consequence (impact)* of the risk. Risks with the highest *likelihood* of occurrence and the highest *consequence (impact)* to the end-product manufacturer score a possible 25 out of 25 points.

There are three main areas in a heat map:

RISK MAP LIKELIHOOD DETERMINATION		
Level		**Approach and Process**
5	Near Certainty	Cannot mitigate this type of risk; no known processes; or workarounds are available.
4	Highly Likely	Cannot mitigate this risk, but a different approach might.
3	Likely	May mitigate this risk, but workarounds will be required.
2	Low Likelihood	Has usually mitigated this type of risk with minimal oversight in similar cases.
1	Not Likely	Will effectively avoid or mitigate this risk based on standard practices.

- **Green area.** Little to no risk. This is in the lower left-hand corner of Figure 22. If printed in color, its color is green and is a whiter shade of gray in the black and white figure. The likelihood and consequence area has 1 to 4 risk levels.

- **Yellow area.** Medium risk. This is in the middle of the chart. If printed in color, its color is yellow and is a darker shade of gray in the black and white figure. The likelihood and consequence area has 5 to 12 risk levels.

- **Red area.** High risk. This is in the upper right-hand corner of the figure. If printed in color, its color is red and is a dark shade of gray in black and white as shown in Figure 22. The likelihood and consequence area has 16 to 25 risk levels.

RISK MAP CONSEQUENCE DETERMINATION			
Level	Technology	Process	People
5	Unacceptable performance, no alternatives exist	Ad hoc work. No plans. No controls. No continuity. No scalability.	Lack of mapping of skills to work
4	Unacceptable performance but workarounds available	Additional tasks required. Fewer work arounds. Minimal process charts, procedures, and little work instructions.	Minimal processes and mapping to work.
3	Moderate performance shortfall, workarounds available	Stable and controlled processes. Extensive procedures and work instructions. Work is measured.	People know what to do to get work accomplished consistently.
2	Minor performance shortfall, same approach retained	Extensive processes are stable and capable of meeting requirements. Work metrics are achieved.	High level of proceduralization and standardization of work. Strong engagement.
1	Minimal Impact	Process management and improvement are extensive. Parts per million defect rates. Enterprise, programmatic, transactional milestones reached.	Extensive training and development applied to all enterprise processes. Self-managed teams.

LIKELIHOOD AND CONSEQUENCE RISK DETERMINATION

As discussed, the risk map construction and evaluation is largely qualitative. To ensure consistency in the heat maps, the end-product manufacturer develops its own or follows the Risk Map Likelihood Determination and Risk Map Consequence Determinations shown in the previous tables.

The Risk Map Likelihood Determination table breaks out likelihood levels 1 through 5. Likelihood determination is probably more difficult to estimate since it is based on assumptions.

Once risk likelihood and consequence are determined, they are plotted on the risk map shown on Figure 22. The above discussion brings up an important topic regarding people. People risks may be the highest risk in SCRM. Why? People are different. They have their methods to do things. Attitudes and education are different. The result is possibly more variation and hence more risk.

PEOPLE, PROCESS, TECHNOLOGY RISK ANALYSIS

The Risk Map Consequence Determination table illustrates risk consequence for

- People risk.

- Process risk.

- Technology risk.

People Risks

SCRM processes are largely people dependent. They can create a high degree of variability and risk in SCRM delivery processes and control systems. For example, the lack of detailed policies, procedures, work instructions can lead to a high level of discretion and possibly risk at the individual problem solving (RBPS) and decision making (RBDM) level.

In the case of well-trained, well-intentioned employees, a key-person syndrome can be created. Operations managers may become indispensable. The challenge is people dependent processes become choke (risk) points, when a key-person leaves. This is exacerbated when there are no succession plans to back-up or replace current key SCRM managers who could soon reasonably retire. Also in the case of poorly trained or less well-intentioned employees, there is an elevated potential for error, fraud, or abuse in the system.

Institutional SCRM knowledge often rests with a key person, not the position. The key-person risk can result in lack of institutional knowledge, lack of risk-control, lack of continuity, lack of consistency, etc. People strains and risks can reside in SCRM processes, such as engineering, purchasing, contracting, human resources, etc.

Process Risks

Business continuity and consistency are key indicators of process controls, strong institutional memory, and process/function dependent controls instead of person dependent controls. In most organizations, key SCRM work presently flows vertically and horizontally through the end-product manufacturer. As the trend to more process based work continues, there will be continued challenges to develop internal process, risk-controls.

Key SCRM processes can be spread across divisions, projects, and processes that involve suppliers. In a global end-product manufacturer, each division touching SCRM work can lead to less operating efficiency; higher variability; lower customer service; higher risk of miscommunication; misallocation of resources; and hand-off risk among process owners, and scalability risk. This can impede the ability to develop and deploy efficient and scalable SCRM processes and eventually be able to integrate them seamlessly across the supply chains.

Technology Risks

SCRM technology risks are becoming increasingly important. SCRM cyber security risks involve state-of-the-art information technologies that must be secured from cyber-attacks. IT systems and software can be major sources of risk, which will be addressed in another book.

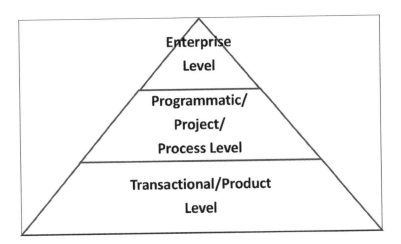

*Figure 23: **Hierarchal Risk Lens***

HIERARCHAL RISK ANALYSIS

Customer-supply risks can be evaluated through hierarchal lens as shown in the above Figure 23. End-product manufacturers can view supply chain risk through:

- Enterprise risk lens.

- Programmatic/Project/Project risk lens.

- Transactional/Product risk lens.

Enterprise Risk Lens

SCRM should start at the enterprise level. This is why SCRM is often referred to as an Enterprise Risk Management (ERM) initiative. The end-product manufacturer may start a SCRM initiative by conducting a risk assessment of the supplier's financial capability, business model, opportunity risks, Tone at the Top, ERM program, and compliance.

SCRM looks at the critical risk elements of the supply chain not only in terms of what may impede production of goods and services but also at critical issues that can impact the company's reputation such as prison or child labor used by suppliers.

PROGRAMMATIC/PROJECT/PROCESS RISK LENS

Programmatic, project, or process lens looks at different SCRM indicators. A programmatic or project risk assessment may look at the supplier's cost, schedule, scope, and quality risks.

There are many SCRM project risks with related impacts and consequences. Most deal with some variance (risk) from a defined target (objective). A project's quality, cost, or schedule may be off target, then each variation is a risk to project success. Each variation off target is a possible risk that must be controlled:

- **Project quality risk.** Supplier project quality variances are controlled through supply selection, supply development, SPC, risk assessment, product testing, detailed specifications, SLA's, or incoming inspection.

- **Project cost risk.** Supplier project cost variances are corrected through design reviews, training, review of work, review of change orders, certification, or other mitigations.

- **Project schedule risk.** Supplier schedule cost variances may involve: early shipments, late shipments, wrong location of shipment, incorrect or inappropriate carriers, or incorrect product sequencing. These can be controlled with the selection of the right carrier, detailed instructions, real-time delivery monitoring, etc.

A process lens looks at the supplier's processes and ability to manage and comply with the end-product manufacturer's expectations, requirements, and specifications (upstream focus).

Looking at a customer-supply problem through the process lens, the end-product manufacturer may request Statistical Process Control (SPC) charts of the process that made the nonconforming products and may even audit the supplier (RBPS).

The process lens can be used to see how well, the supplier manages its own suppliers. In other words, a process analysis would look down the supply chain in terms of how first-tier supplier manages its second and even third-tier suppliers. The analysis would look at the risk-controls for managing suppliers, process capability controls, and even supply project controls.

TRANSACTIONAL/PRODUCT RISK LENS

Product risk lens looks at product attributes or value characteristic that do not meet specifications or comply with regulations. Product risk lens may involve analyzing critical product dimensions, labeling, color consistency, and other value attributes. Transactional attributes may include delivery times or service level agreements.

If there is a problem with product delivery or a product nonconformance, then the end-product manufacturer will contain the nonconforming shipment and conduct 100% or statistical inspection of incoming shipments (RBPS).

ENTERPRISE RISK APPETITE

One of the major challenges with SCRM is the end-product manufacturer needs to define its risk appetite at the enterprise level and then deploy risk tolerance to programs and lower organizational levels. The following quote describes the challenge:

> "Designing risk management without defining your risk appetite is like designing a bridge without knowing which river it needs to span. Your bridge will be too long or too short, too high or too low, and certainly not the best solution to cross the river in question."[101]

Risk appetite is defined at the Board level for the enterprise because it is linked to the overall strategy and risk profile of the end-product manufacturer. Factors to consider in developing an enterprise level view of SCRM risk include:

- Solvability, liquidity, earnings, and earnings volatility.

- Credit rating.

- Reputation and brand.

- Expansion into new product, end-product manufacturer groups.

- Supply chain risk management.

- Acquisitions.

- Environmental impact.

- Corporate governance and compliance.

- Human resources.[102]

Once the enterprise risk appetite has been defined, then lower levels of risk tolerance can be defined at the Programmatic/Project/Process level and Transactional/Product levels. The end-product manufacturer should negotiate risk tolerance with key suppliers based on defined and mutually acceptable risk targets and risk limits with the supplier.

The risk target can be a SCRM objective. The risk limits become the tolerance thresholds to monitor with the supplier, which may include cost, schedule, and quality variances. Quality variances may include reliability goals and production/process controls based on Statistical Process Control limits. Schedule variances may be delivery within a two-day early window and a one-day late window. The early window implies the truck is unloaded and supplied material is placed in production or incoming inventory. The late window implies there is sufficient stock to maintain production and if the product is delivered beyond the window, then production stops. Much in the same way, risk tolerance can be defined for cost, technology, scope, and other critical factors.

RISK ANALYSIS OUTPUT

Once the risks, consequences, and probability of the risks occurring are understood, the SCRM team can develop an overall supply chain risk profile for a product segment, then break down the risk profile for individual suppliers within the segment. The profiles prioritize risk areas and variables from highest to lowest. Using the Pareto Principle, the operations manager knows to focus his or her attention on the suppliers with the highest risk and design SCRM controls to minimize risks.

CONTEXT: When Should Risks be Quantified?

- High likelihood or probability of risk occurrence or exposure.
- High consequence or impact of the risk occurrence or exposure.
- Life, safety, or environmentally threatening event.
- Final-customer or market condition changes.
- Major changes in the input variables in a supply chain process, including changes in personnel, management, environment, materials, suppliers, etc.
- Process instability or lack of supplier capability of meeting requirements

---

TREAT (MANAGE) RISKS

Less is more. God is in the details.
Mies van der Rohe

We have discussed two definitions of risk. Upside risk involves searching for opportunities for investment and downside risk involves mitigating high consequence and high likelihood risk.

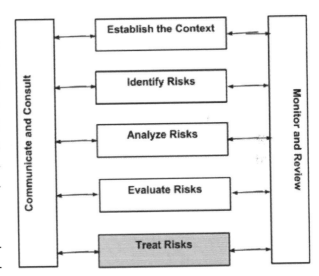

In this section, we discuss how to 'treat (manage) risks' within the end-product manufac-

Figure 24: ISO 31000 - Treat Risks

turer's risk appetite and tolerance. Once risks are identified, risks can then be mitigated.

Risk treatment and mitigation plans are always designed and tailored to the end-product manufacturer's context, business model, regulatory environment, and products. Risk plans identify internal controls, risk treatment, and processes for reducing risk to within the risk appetite of the end-product manufacturer.

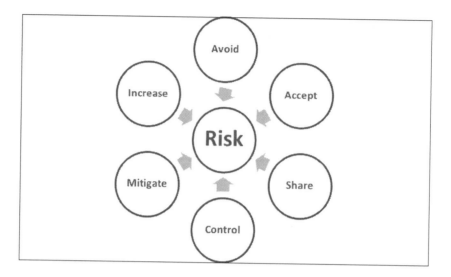

Figure 25: *Risk Mitigation Strategies*

The following are common risk treatment strategies:

- Avoid risk.

- Accept risk.

- Increase risk.

- Share risk.

- Control risk.

- Mitigate risk.

The following examples of risk treatment can fall within multiple treatment categories. Such as, a risk sharing example can be used to mitigate risks and vice versa.

AVOID RISK

Risk avoidance is the choice to stop, avoid, or discontinue supply management activities that can cause a risk or can result in an increase in the likelihood and/or consequence of the risk. Risk avoidance is usually determined after a risk assessment (RBPS). The assessment determines that new product development will not return the investment (ROI) required by the end-product manufacturer. The assessment determines the risk of new product development is higher than the risk appetite or tolerance of the end-product manufacturer. Or, the end-product manufacturer determines the product is too complex and the current supplier cannot design and produce it at the cost required and determines to make the product internally.

Examples of risk avoidance include:

- Decide that a project is not worth the investment. The option is to cancel the project.

- Decide a supplier cannot produce the product or service to end-product manufacturer requirements. The option is to select another supplier or find a more responsive or capable supplier that can meet the end-product manufacturer's requirements.

- Use less complex technologies or methodologies if possible with commodity suppliers.

- Use stable designs with suppliers and follow a change management process with product development updates or changes in specifications.

- Review multimodal transportation and distribution options at critical supply chain hubs or nodes.

- Develop common electronic communication protocols with key suppliers.

- Encrypt crucial customer-supply communications such as designs, Intellectual Property, contracts, and other information, so proprietary ideas cannot be stolen.

- Standardize end-product manufacturer SCRM processes, guidelines, protocols, specifications, and requirements, so the supplier can provide compliant products or services.

- Centralize critical activities with appropriate fallbacks and backups. If there is a supply disruption, then the backups can keep critical supply chain operations operating.

- Move critical suppliers up the capability and maturity curve.

CONTEXT: Supply Chain Value Management Tips

- **Plan long-term.** Choose a product and offer a solution, that can accommodate to changes throughout the product lifecycle.
- **Ensure short-term results.** Focus on short SCRM projects and obtain demonstrable results that matter.
- **Qualify suppliers.** Certify and eliminate suppliers (RBPS) with inventory and customer service problems.
- **Be aware of vulnerabilities in the supply chain.** Some points or nodes in the supply chain are more critical than others. Points of vulnerability may be politically unstable countries, insecure transportation methods, or vulnerable distribution centers.
- **Be aware of key supply chain decision points.** Many organizations have not identified key decision or risk-control points (RBDM) in their supply chain systems and processes.
- **Focus on one piece of the supply chain puzzle at a time.** There are many elements, links, and activities in a supply chain. A company is probably part of someone's supply chain as well as has its own supply chain (RBDM).
- **Partner with or hire logistics talent.** SCRM solutions require specialized knowledge and assistance. Invest in getting and institutionalizing specialized knowledge and skills (RBPS).
- **Do not underestimate the difficulty of connecting to a host or other computer systems.** SCRM solutions usually integrate into many platforms, including legacy systems.
- **Consider the wider SCRM community.** SCRM solutions affect end-product manufacturers, suppliers and freight companies. Make sure everyone understands SCRM goals and deliverables.
- **Source multiple suppliers if there are risks.** Do not put all your eggs in one basket. Diversify supply chain risks with multiple suppliers of critical products (RBDM).
- **Focus on supply development.** Educate suppliers on new KRI's and train them how to use them to achieve end-product manufacturer business goals.
- **Develop suppliers carefully.** Look for and integrate cutting-edge, synergistic supplier competencies.

ACCEPT RISK

This means the end-product manufacturer has identified the risk, its consequences, and can live with them. The operations manager accepts the fact that shipments from a supplier may have an acceptable number of nonconforming products. If the quality level of received products needs to be improved then the end-product manufacturer or supplier can invest in additional machinery to improve capability levels.

Accepting higher risk is sometimes counter intuitive. If SCRM professionals accept higher risks, then possibly there are more heartaches, but the end-product manufacturer may accept these. For example, if the benefits of high risk lead to much higher gains, then risk acceptance is worth it. For example, specifications for a product or service attribute may be loosened, so the product is made at a lower cost.

Or, a company may use its value-added differentiators to accept or even increase risks to take advantage of market conditions and the company's competitive differentiation such as 3M Corporation:

> "One area where 3M has a high (risk) tolerance is on a macroeconomic or geopolitical level because of this idea that experience reduces risk volatility. We have been in a lot of these volatile markets for many years – Brazil for more than 70 years, we were one of the first multinationals to go into China more than 30 years ago and we have spent a similar amount time in Turkey. Companies we know that tend to do well in these markets think more longer-term, think in terms of quarter-centuries, not financial quarters."[103]

Examples of risk acceptance include:

- Pursue SCRM projects that fall within the risk sensitivity or tolerance of an activity, i.e.
    - Fall within the risk appetite of the enterprise or end use manufacturer.
    - Fall within the project risk tolerance for scope, cost, delivery schedule, or quality.
    - Fall within the service level required or the specification limits of the product.

- 'Make' not 'buy' critical products and services.

- Purchase products or services from certified suppliers, that pose a lower overall risk.

- Purchase non-critical or commodity services/products from smaller or uncertified supplier.

- Use commodity products for noncritical applications.

- Lower the level of requisite risk-assurance with a supplier.

- Reprice product/service offerings to make them more attractive to buyer or consumer.

- Offset risk by special pricing, 24/7 availability, and delivery etc.

- Self-insure if possible against supplier risks.

INCREASE RISK

This means the end-product manufacturer will increase its risk appetite or tolerance because the potential returns outweigh the risks. Increase risk is often similar to accept risk. For example, an end-product manufacturer may select a new supplier with proprietary technology knowing that its processes are immature and prone to failing. Why? The future potential returns outweigh present perceived risks.

Examples of increasing risk include:

- Change the risk sensitivity at the:
  - Enterprise level.
  - Programmatic/Project/Process level.
  - Transactional/Product level.

- Determine the supplied commodity product is used for a non-critical application.

- Conduct risk benefit analysis.

- Determine using the supplier will generate returns substantially above the cost of capital.

SHARE RISK

This means a risk is unacceptable and is spread to lower an individual risk. The operations manager has a sole-source supplier and the risk of an event such as a strike implies that shipments may be disrupted. The end-product manufacturer may spread or diversify the risk by finding acceptable, alternate suppliers.

Risk sharing is a hedging strategy used by sectoral companies. For example, electric power companies will jointly purchase a costly transformer and keep it in storage so that if one of the utilities has a transformer problem then the transformer in inventory is used for the emergency. The cost of storage and purchase is shared among the utilities.

Insurance is another tool for sharing risk in global supply chains. Insurance is a form of sharing risk with a third-party. Insurance provides a method to managing supply chain interruptions such as the lack of critical material, natural catastrophe, cyber breach, or plant shut down. Insurance is often a transactional risk management tool that is highly effective.

Insurance however has its limitations especially as protection against critical supply chain disruptions. Insurance is not a substitute for good SCRM and RBDM. Insurance usually cannot account for reputational loss, share price reduction, or loss of market share. Also, insurance agreements do not typically address political risk, local regulations, protectionism, local competition, consequential damages, reputational risk, or poor decisions (RBDM).

Examples of risk sharing include:

- Outsource non-core work to suppliers.

- Require suppliers to be ISO 9001:2015 certified.

- Purchase cyber security insurance.

- Conduct second and/or third-party Quality Management System audits.

- Conduct risk-assurance audits of critical suppliers.

- Use QA/QC third-party service provider to evaluate the quality, schedule, and critical supplier shipments.

- Use accredited third-party laboratories to test supplied products.

- Retain prime and acceptable alternate suppliers of critical equipment and services.

- Back up critical data and information daily.

- Purchase Errors/Omission insurance for professional audits of critical suppliers.

- Indemnify third-party consultants in critical areas, such as cyber security, artificial intelligence, robotics, and risk-assurance.

- Review and securitize joint assets with suppliers.

- Design and deploy appropriate supplier risk-assurance based on risk tolerance for each supplier and product.

- Reinsure part or all of critical supplier risks for being late on arrival or for nonconforming products.

- Retain prime and acceptable alternate suppliers of critical products and services.

- Share risks with suppliers for variability in delivery, cost increases, testing of products, or containment of nonconforming products.

- Maintain alternative sources of supply for critical products and services.

- Decrease the probability of disruption (spread the risk).

- Secure alternative routing, carriers, and modes in advance.

- Maintain incoming, in process, and final inventory of critical products.

- Build supply chain redundancies at critical supply chain decision points.

- Develop alternative port, logistics, and transportation strategies for critical outsourced products.

- Use of labeled and modified containers based on containment of the hazardous material (RBDM).[104]

CONTROL RISK

This means the occurrence or recurrence of the risk is monitored and even prevented. The operations manager has sufficient supplier trend data or other product information to predict when a shipment may contain nonconforming products, when a machine's output is out of specification, or when machines are not preventively maintained.

End-product manufacturers want to control supply risks so they can be reasonably assured that business and SCRM objectives can be met. There are important implications about controlling supply chain risks:

- Risk-control is a process. It is a means to an SCRM end, not an end in itself.

- Risk-control is conducted by people. It is not merely policy manuals and forms, but people at every level of the supply chain are responsible for their work and controlling risks.

- Risk-control is expected to provide only *reasonable assurance,* not absolute assurance of supply chain risks to executive management.

- Risk-control is geared to the achievement of SCRM objectives in internal and overlapping SCRM processes.[105]

Examples of risk-control include:

- Lower level of overall enterprise risks.

- Design and deploy supplier or third-party risk-controls at:

    o   Enterprise level.
    o   Programmatic/Project/Process.
    o   Product/Transactional.

- Increase level and breath of customer-supply risk-controls.

- Apply specific enterprise, process, and project level SCRM methodologies with different suppliers.

- Use redundant SCRM systems with critical suppliers.

- Use joint application design (JAD) in enterprise risk software development.

- Use design-build practices in managing the design and construction of internal facilities.

- Add schedule float to SCRM project schedules.

- Develop process controls for critical product quality attributes with appropriate capability indices.

- Identify critical product attributes for supplied products and develop quality controls for these attributes.

- Select and certify suppliers for similar products and services based on common product value attributes.

- Develop suppliers using improvement technologies such as lean, Six Sigma, operational excellence, etc.

- Monitor supplier performance using KPI's and KRI's.

- Develop segment-specific, supplier process/product risk management methodologies.

- Automate and standardize key SCRM processes.

MITIGATE RISK

This requires the probability or likelihood of an event occurring is reduced. As well, the severity or consequence of an event is reduced. Risk mitigation may involve either or both of these techniques.

Being proactive, predictive, pre-emptive, and preemptive (4P's) allows an end-product manufacturer to respond quickly to unforeseen events and be able to meet critical business and SCRM objectives. In this way, the company can focus on identifying and mitigating risks that inhibit meeting a business or SCRM objective.

Risk likelihood reduction examples include:

- Require equivalent risk-controls from critical suppliers.

- Define supplier risk-controls and supplier capability requirements.

- Develop appropriate people, process, and technology controls at:

  o Enterprise level.
  o Programmatic/Project/Process.
  o Transactional/Product.

- Modify customer-supply controls at:

  o Enterprise level.
  o Programmatic/Project/Process.
  o Transactional/Product.

- Develop precise and accurate standards for products and services.

- Improve contracts to detail specific quality, cost, schedule, technology, and software requirements.

- Review and update product specifications with appropriate testing of product quality attributes.

- Require supplier certification and improvement of suppliers of critical products and services.

- Develop supply contingency plans based on 'what if' questions.

- Develop internal SCRM controls.

- Develop internal project management controls.

- Require RBPS and RBDM controls to be deployed by critical suppliers with evidence provided to end-product manufacturer.

- Require supplier preventive maintenance on supplier machinery.

- Require Product Part Approval Process of critical supplier machinery and processes.

- Develop internal quality management and assurance practices.

- Deploy internal and supply SCRM training.

- Train internal operational management supervision on risk-controls, 3 lines of defense, risk monitoring and risk-assurance.

- Develop new SCRM organizational structure with additional authorities and responsibilities.

Risk consequence reduction examples include:

- Develop new contracts that are more favorable to the end-product manufacturer.

- Conduct business impact and continuity analyses.

- Develop contingency plans if the supply chain is disrupted.

- Develop disaster or incident recovery plans if there is a supply disruption

- Develop better product designs with detailed specifications, drawings, and requirements.

- Scope work and deliverables accurately.

CONTEXT: Risk Management Questions

- What risks can the end-product manufacturer accept?
- What risk management strategy does the end-product manufacturer want to follow?
- What role does the end-product manufacture want the operations manager to assume?
- What are specific system, process, or product risks?
- What is done to mitigate each risk and control critical process variables?
- What treatment is deployed, i.e. risk transfer, risk acceptance, risk reduction?
- What happens if risks are accepted?
- What type of contingency plans are developed?

- Add incoming, in process, or final inventory to anticipate supply deficiencies.

- Add production or service capacity with existing or new suppliers.

- Develop approval and review gates in SCRM projects and contracts.

- Minimize exposure to risk sources by developing SCRM plans.

- Develop internal cost controls for suppliers.

- Back up data and critical information daily.

- Conduct additional SCRM planning for suppliers with delinquent histories.

- Conduct supplier Quality Management System or forensic audits of critical suppliers.

MONITOR AND REVIEW

Tell me and I will forget, show me and I may remember, involve me and I'll understand.
Chinese Proverb

Supplier Monitoring and Review closes the architect, design, deploy, and assure cycle.

The following are commonly used techniques to monitor a supplier:

- Control environment.

- Audits.

- Product samples.

- Process capability analysis.

- Incoming material inspection.

- First-article inspection.

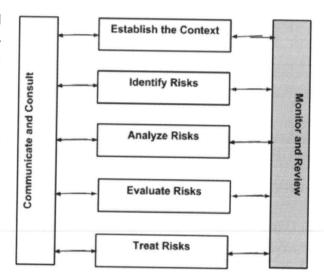

Figure 26: *ISO 31000 – Monitor and Review*

CONTROL ENVIRONMENT

SCRM enterprise lens looks at developing an appropriate governance and risk-control environment based on the context of the end-product manufacturer. Sometimes, this is called SCRM governance. The Board of Directors provides guidance to executive management regarding end-product manufacturer goals, business model, and risk appetite. Executive management's objective is to provide direction to the end-product manufacturer in terms of SCRM design, deployment, and assurance of risk-controls. The end-product manufacturer may require key suppliers to have similar control environment, including ERM, SCRM, RBPS, and RBDM as it has.

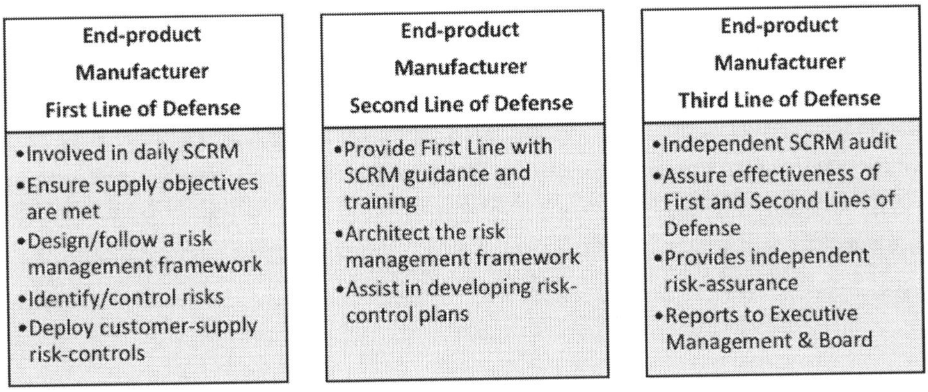

End-product Manufacturer First Line of Defense	End-product Manufacturer Second Line of Defense	End-product Manufacturer Third Line of Defense
•Involved in daily SCRM •Ensure supply objectives are met •Design/follow a risk management framework •Identify/control risks •Deploy customer-supply risk-controls	•Provide First Line with SCRM guidance and training •Architect the risk management framework •Assist in developing risk-control plans	•Independent SCRM audit •Assure effectiveness of First and Second Lines of Defense •Provides independent risk-assurance •Reports to Executive Management & Board

Figure 27: **Three Lines of Defense**

More often, the end-product manufacturer may design three lines of defense regarding the supply control environment as shown in the above Figure 27. The first line of defense consists of supplier process and project owners who own their risk-controls. Key supplier process owners are empowered to make RBPS and RBDM to ensure key customer-supply SCRM objectives can be met.

The second line of defense includes the end-product manufacturer SCRM process owners, who ensure customer-supply controls and risk-assurance are working effectively. The end-product manufacturer identifies SCRM objectives, constraints to meeting the objectives and manages these constraints or risks across the into the supply chain. The second line of defense can include SCRM risk management, risk-assurance, quality assurance, compliance and even product inspection. This line of defense monitors, trains, and facilitates the deployment of SCRM into operational management and into the supply chain. This group also facilitates reporting this information to relevant stakeholders within the end-product manufacturer.

The third line of defense involves independent assurance functions within the end-product manufacturer. The third line may include internal auditing, quality auditing, supplier surveillance, supplier auditing, ISO Certification Bodies, and contractual reviews. The third line of defense may report SCRM risks to the Board of Directors and determine how well customer-supply relationships are being managed. As well, the third line of defense provides independent risk-assurance of the effectiveness of the first and second lines of defense.

AUDITS

There are 3 basic types of customer-supply audits: 1. System, 2. Process, and 3. Product. ISO 9001:2015 management system audit checks internal documentation for compliance and the supplier's ability to comply with ISO standards. A process audit checks the supplier's cost, quality, delivery, and other critical processes for stability, capability, and innovation. A product audit checks the supplied product for conformance to technical standards, performance standards, etc.

A team representing quality, manufacturing, engineering, and supply management may conduct critical customer-supply audits. Major suppliers are those that provide a large dollar volume of products or supply key components. A major benefit for the adoption of ISO 9001:2015 standards is the ability to check a register of certified suppliers to avoid duplicative audits.

Audits are usually formal and highly structured. They begin with an initial discussion about the purpose, scope, and intent of the audit. Depending on the type, intent, and scope of the audit, the following areas are investigated:

- Quality manual, procedures, and work instructions.

- Organizational structure.

- Logistics processes.

- Cost sharing.

- Technical capabilities.

- Web or business to business capabilities.

- Training and certifications.

- Documentation.

- Final product test and evaluation.

- Corrective/prevention actions.

- Measuring equipment and calibration.

- Storage and delivery.

PRODUCT SAMPLES

End-product manufacturer may send a sample product to a supplier when specifications are difficult to develop, such as for products with intricate shapes. Samples are also sent to suppliers to determine if they can be manufactured to specifications and tolerances such as where the standard value is 0.500 inch and the upper and lower specification limits are respectively 0.510 and 0.490 inches. This is illustrated in Figure 28 on the next page.

Sometimes, parts cannot be specified in writing or depicted in drawings because of intricate geometric shapes. Instead, it is easier to produce a sample product and give this to a supplier to duplicate. The supplier then makes a similar product and submits this to the end-product manufacturer with critical dimensions clearly marked on engineering drawings. Then, duplicate measurements using similar instruments are made by both supplier and end-product manufacturer. Appearance, critical measurements, and product performance are verified and any differences are checked. The problem with samples is that product reliability cannot be tested. So, a product may conform to all specified measurements but not survive prolonged use (RBPS).

PROCESS CAPABILITY ANALYSIS

Process capability analysis is a statistical study of dimensional variation in a product characteristic. The end-product manufacturer provides the supplier with a few parts. Dimensions on parts are measured and compared against the specification. Deviations between the actual measurements and the specifications are analyzed. If measurements are inside and centered in the middle of the specification spread, products are accepted for use. If measurements are outside the specification limits, products are returned to the supplier.

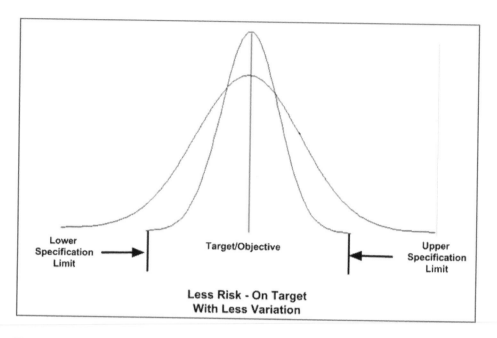

Figure 28: *Process Capability*

INCOMING MATERIAL INSPECTION

Incoming material inspection measures, tests, or checks material from suppliers. Incoming material inspection is time consuming and costly. By checking for product nonconformances, the end-product manufacturer is ensuring the quality of the supplier's work. Material is accepted, rejected, or corrected. If accepted, material goes into inventory or directly into production. If rejected, material is used 'as is', reworked, or returned to the supplier.

FIRST ARTICLE INSPECTION

ISO standards define inspection as the "examination of a product, process, service, or installation or their design and determination of its conformity with specific requirements, or on the basis of professional judgement, with general requirements." [106]

First article inspection is used for inspecting, testing, and measuring the first of a series of products or a significantly modified product that is manufactured by a new or existing supplier. First article inspection is a cursory check of critical product or service characteristics or a comprehensive investigation of all physical, functional, and dimensional characteristics of the part or assembly. The degree of the investigation depends on the complexity of the product, familiarity of the supplier, and time constraints.

First article inspection can:

- Detect early discrepancies, defects, or flaws in the prototype product.

- Serve as proof that specifications are met.

- Serve as proof that manufacturing set-up is correct.

- Confirm that corrective action modifications have been made.

However, first article inspection can provide misleading results because it:

- Is a snapshot of one item and does not indicate the conformance of the entire lot or batch of production products.

- Does not indicate product reliability.

- Does not allow for any process variation that may move dimensions out of the acceptable specification spread.

RISK MANAGEMENT FRAMEWORK THOUGHTS

In god we trust; all others bring data.
W. Edwards Deming

'Where to start' is always a critical question? If the operations manager is dealing with high-value, technology, process, or quality content then these would be managed more carefully than commodity or fixed price items.

STARTING SCRM ARCHITECTURE
The end-product manufacturer should focus on understanding, analyzing, and designing critical internal SCRM processes. Each process has process variables, constraints, RBPS, and decision points. Each of these should be assessed for risk.

What does the operations manager do if he or she does not have sufficient process knowledge or information to identify risks? The operations manager should get qualified experts or preferably process owners to identify process risks and controls. Following this process through each customer-supply tier, the end-product manufacturer and its suppliers can mitigate critical risks in the supply chain.

LAW OF UNEXPECTED CONSEQUENCES

You get what you ask for in SCRM. If SCRM if architected, designed, deployed, and assured using a risk management framework, there is a higher level of probability and assurance that SCRM objectives can be met and risk-assured.

However, there is never absolute assurance. The end-product manufacturer should be aware of the law of unintended consequences. Let us look at automotive airbags. Their purpose was to save lives in the event of an auto accident. However, the Takata massive global recall of airbags demonstrated the law of unintended consequences. The Takata massive global recall started when automotive manufacturers required drastic cost reductions from Takata. Takata responded by using ammonium nitrate in its airbags in 2001. Ammonium nitrate was 1/10 of the cost of the prior compound, but it was inherently unstable when it burst resulting in the largest recall in auto history.[107] Takata certainly did not expect the airbags to explode and cause injuries.

Following this process through each customer-supply tier can mitigate critical risks supply chain risks. Again, this is not providing absolute assurance but providing reasonable assurance hopefully within the risk appetite of the end-product manufacturer.

SUPPLY CHAIN COMPLEXITY

Supply chains by their very nature are complex involving many companies, distribution centers, warehouses, production facilities and stakeholders. Over the last 5 or so years, food, pharmaceuticals, electric power, electronics, and many sectors have developed their own SCRM requirements and specifications.

Simplification is a key risk based approach to eliminate programmatic, pro-ject, process, and product complexity. These complexities are risks that may result in unexpected supply chain disruption, increase delivery sched-ule, or increase time-to-market product development.

SCRM IS COUNTERINTUITIVE

SCRM is often counterintuitive. SCRM risks may come from unexpected sources or events. For example, it is generally believed the highest risk in the supply chain may reside with suppliers with the highest spend. How-ever, this may not be the case. Highest risk may reside with small suppliers that have a proprietary product or service or a single-source supplier of a critical product. This is one of the critical unknowns of SCRM.

The 80-20 rule is the Pareto Principle. The Pareto Principle is a simple example of RBDM. Expressed in supply chain terms, the Pareto Principle says that 20% of the suppliers will cause 80% of the problems. Thus, the Principle identifies the most critical suppliers to risk-control. Or expressed another way, the Pareto Principle can help identify the most critical supply chain variables that pose the highest risk of failing, not meeting require-ments, or causing problems.

ISO 31000 TAILORING CHALLENGES

ISO 31000 is generic and descriptive. The standard offers flexibility and adaptability. The standard must be architected and designed depending upon final-customer and end-product manufacturer requirements. Unfor-tunately, there is no prescriptive method for architecting, designing, deploy-ing, and assuring the SCRM framework.

ISO/TR 31004: Risk Management – Guidance for the Implementation of ISO 31000 provides the following guidance for tailoring the risk manage-ment framework to an organization.

- Global end-product manufacturers may have different SCRM con-texts depending on services, products, final-customers, and stake-holders. Each of which may require a modified and tailored SCRM framework for design and deployment.

- External consultants may not understand the end-product manufacturer's or supplier's contextual requirements, needs, and expectations so there may be a long learning curve.

- End-product manufacturers may be using an existing framework such as COSO that can be adapted to ISO 31000 SCRM.

- Different interpretations may exist of the level, extent, and nature of the SCRM risks faced by the end-product manufacturer.

- Supply chain risks are sometimes unknown and unknowable.

- Supply chain risks interact in unexpected ways.

- Multiple supply variables can have unknown consequences or impacts.

- Single risk variables can interact with others that result in multiple unexpected effects.

- End-product manufacturers, process owners, and suppliers may be too close to the problem to conduct SCRM RBPS and RBDM.

- SCRM mathematical modeling can provide a false sense of security and accuracy.[108]

SUMMARY

SCRM using ISO 31000 is not easy or straight forward. It requires deep and broad knowledge of the end-product manufacturer and the supply base. It requires a high level of customer-supply trust. It requires deep and broad knowledge of the standard. It requires the ability to tailor the architecture, design, deployment, and assurance of the standard to the specific context of the end-product manufacturer, supply base, and products.

Supplier trust is also critical. Suppliers may be reluctant to share proprietary cost, margin, and performance information with end-product manufacturers. This makes sense. If a supplier is selling a product at a 30% margin to one end-product manufacturer and only 25% to another, the supplier does not want to share this sensitive information for the obvious reason.

The end-product manufacturer paying the higher amount will demand the supplier match the lower price.

NEXT CHAPTER

In the next chapter, we discuss SCRM leadership and teaming, which are critical elements of a successful SCRM initiative.

CHAPTER 8: SCRM DEPLOYMENT - LEADERSHIP

WHAT IS THE KEY IDEA IN THIS CHAPTER?
SCRM deployment requires a culture change. People by their nature do not like change. It is personally challenging. It can be painful. It requires learning new behaviors. It requires doing things differently. And, there is no guarantee of success.

Studies as well as our experience indicate that leadership is the #1 factor of SCRM success.

BOUNDARY LESS ORGANIZATION
Jack Welch, the CEO of General Electric, popularized the concept of the boundary less organization where artificial boundaries to resources, communication and cooperation are removed to improve business processes and develop new business models.

VUCA technology and instant communication are accelerating the redesign of organizational processes through the removal of horizontal barriers within the end-product manufacturer, the vertical barriers in the organizational hierarchy, and the external barriers between the end-product manufacturer and supply base.

What we do know about SCRM can be distilled into one premise: VUCA will accelerate SCRM disruption. Hopefully, institutions will be resilient. Business models will be adaptable. People can adopt new behaviors.

We can probably agree that VUCA is the new normal. But, we believe that many institutions, companies, and people are not ready.

In this chapter, we discuss how leaders, manager, and people may have to assume new SCRM roles and learn new skills to help the organization be VUCA-ready.

ORGANIZATIONAL TRANSFORMATION

Two basic rules of life are: 1) Change is inevitable. 2) Everybody resists change.
W. Edwards Deming

The best designed and well-intentioned SCRM initiatives will fail to deliver business performance unless the organizational and human dimensions of implementation are addressed early on. Any SCRM organizational change, incremental or transformative, is immensely difficult to accomplish as SCRM executives want VUCA-ready, supply chains.

CHANGE MANAGEMENT

The SCRM journey is a business transformation. Business transformations follow a value and process based approach to change. Names and fads may change but the evolution is usually the same.

As a result of competitive and disruptive pressures, end-product manufacturers have to question what they are doing, why they are doing it, and how they can do it better. Nowadays, entire industry sectors are being transformed or disrupted by competitive VUCA pressures.

Some highly adaptable end-product manufacturers are more conducive to change than others. In our experience, mature end-product manufacturers in commodity or regulated sectors are less likely to change because of entrenched practices, top heavy hierarchies, and inflexible cultures. Continuous innovation and innovation tend to flourish in end-product manufacturers and sectors where the marketplace requires speed, agility, and flexibility. Supply chain systems and processes by necessity must be flexible to adapt to a dynamic marketplace where final-customers want 'built to order' products. Supply chain processes must also be lean, congruent, supportive, SCRM focused, and reinforce the culture of RBT, RBPS, and RBDM.

CONTEXT: Supply Management Challenges

- Supply management organizations are sometimes perceived as cost centers.
- Supply management organizations have a transactional or product focused view of purchasing.
- Operations managers on the verge of retiring are placed in supply management until they retire.
- Managers and employees do not understand the full breadth of global supply management. We still see that 'Negotiation 101' is the most popular course in purchasing and supply management.
- Managers do not understand how products are designed and manufactured.
- Managers do not understand how services are designed and provided.
- Supply key performance indicators are missing or are not linked to the end-product manufacturer's key strategies and KRI's.
- Supply management department does not have reliable and demonstrable KPI's or KRI's. Or, the indicators focus on transactional metrics/objectives, such as the number of suppliers on boarded and inspection levels.
- Supply management is low on the risk capability and maturity curve.

The most difficult part of deploying a new management concept is getting the end-product manufacturer, stakeholders, and suppliers to understand, accept, and embrace the need for disruption and SCRM change. Getting the support, commitment, and understanding of executive management is a significant hurdle for end-product manufacturers deploying new management approaches.

SCRM IS NOW A CRITICAL SKILL

SCRM has become a top issue in operational management. SCRM is a critical skill for operations management, supply management, purchasing agents, quality professionals, engineers, and all operational professionals. Supply chain management teams are often cross-functional. Cross-functional team representatives may come from quality, production, purchasing, engineering, planning, and other interested stakeholders.

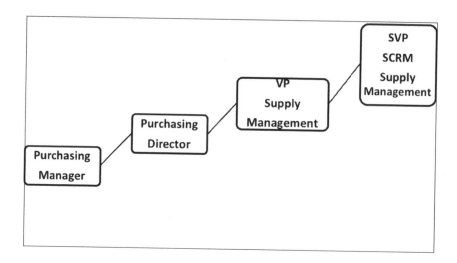

Figure 29: **SCRM Increase of Responsibilities**

Smaller companies still tend to look at supply management as a purchasing function that is cost driven. Operations managers must evolve and adopt new roles. Operations managers have more authority and are more visible than traditional purchasing managers. The profession is also attracting more highly trained and educated personnel. Companies often require operations managers to have engineering and other advanced degrees, professional certifications, and be cross-trained. Experience in product development/design engineering allows supply professionals to challenge supplier's specifications, which are restrictive from a sourcing point of view. For example, operations and design experience allows supply professionals to obtain the respect of engineering and manufacturing personnel while providing input to design and quality decisions (RBDM).

SENIOR EXECUTIVE RISK BACKGROUNDS

The purchasing function until a dozen years ago was a director level function. As more end-product manufacturers developed an outsourcing business model, supply management has been elevated to a senior vice president or even higher position within many end-product manufacturers.

We are also seeing more Chief Supply Management Officers with non-sourcing backgrounds. They bring a new enterprise and process perspective to the SCRM function. They understand risk-control objectives, risk management taxonomies, risk management frameworks, organizational processes, technology integration, and supply chain risk methodologies.

SOURCING CHALLENGES

Architecting, designing, deploying, and assuring SCRM can represent a major organizational challenge. SCRM is often at the enterprise-level. SCRM requires new knowledge skills and abilities. SCRM requires new behaviors of internal managers and from suppliers.

To be successful, SCRM requires executive leadership support. So it is critical that when designing and deploying SCRM, the emotional aspects of the change are addressed especially as they relate to middle management and front-line employees.

SCRM ORGANIZATION

Draw them (organizational charts) in pencil. Never formalize, print, and circulate them.
Robert Townsend, CEO, Avis

What is the best structure to organize the SCRM function?

SEARCHING FOR THE BEST ORGANIZATIONAL ARCHITECTURE

In the last 20 years, several organizational structures have been introduced. These include: traditional-hierarchical, matrix, virtual etc. So is there one best structure? No. The appropriate supply chain structure is one that supports a group of people who can work together to satisfy end-product manufacturers and meet SCRM objectives using the fewest resources possible. Within this structure, congruent, and supportive systems are created that reinforce the SCRM organizational culture.

The traditional form of business organization is hierarchal. However, a supply process chain implies a horizontal, integrated process. Innovative SCRM organizations models are springing up based on interlocking groups, virtual partnerships, strategic alliances, and matrix organizations.

Architecting and designing supplier relationships are critical to ensure efficient, effective, and economic supply risk management. Today, there is more focus on a collaborative arrangement for sharing rewards and risks with critical single-source suppliers. Collaboration may involve joint development of products, sharing information on core SCRM capabilities, internal alignment of business strategies, integration of core manufacturing and other operations capabilities, and mutual agreements on cost, delivery, risk, and technology.

MATRIX ORGANIZATIONS

Matrix organizations became popular to satisfy internal customers and multiple stakeholders. Matrix organizations have been especially popular in aerospace and high-tech supply chain organizations. In the traditional aerospace matrix organization, special SCRM projects usually had multiple reporting structures.

A matrix organization is just what the name implies. A matrix SCRM reporting architecture overlays the traditional functional supply management structure. SCRM teams report up and across the organizational structure. Their work may take them up and down the supply chain. Authority and responsibility for a SCRM project is shared among several groups. This structure works well with supply management teams but has problems when it becomes institutionalized. Several disadvantages to SCRM matrix structures include reporting and responsibility ambiguities, deployment conflicts, and additional costs.

Multidisciplinary SCRM teams also blend well with a matrix organization. The SCRM team may have multiple reporting routes. A supply chain innovation team may report to the supply chain function and to a line organization where the innovation or risk project is conducted. As well, individual multidisciplinary team members may report to different organizational areas.

SCRM teams succeed in a matrix organization because:

- Team leader understands organizational culture.

- Team members understand SCRM requirements.

- Team members are loyal.

- SCRM commitments are made and followed.

- Supplier conflict resolution is fair and quick.

- Good communication channels exist horizontally and vertically.

- Everyone participates in the planning process.

VIRTUAL ORGANIZATIONS

In today's competitive times, key skills are retained through customer-supply partnerships, which result in the growth of virtual organizations. The virtual organization is an amorphous entity that delivers products or services to final-customers with the external appearance of a single company or supply chain.

In much the same way, virtual supply chains will become more prevalent as more small suppliers develop state-of-the-art, world-class competencies, skills, technologies, capabilities, information or other resources. These unique assets are pooled and integrated with other companies to form an integrated network of turnkey supply capabilities. When the supplier contract is over, the virtual supply chain disbands.

Examples of virtual organizations include: virtual supply chains, global consulting organizations, franchising, joint ventures, and strategic alliances. What pulls everyone together in the virtual supply chain is synergy - the awareness that the sum of supplier skills and knowledge is greater than the individual parts.

To make the virtual chain work, supply-partners must bring something of value to the table. It includes special knowledge, special design capabilities, risk management, or other resources. When the project is over, the virtual supply group may develop another product or the group may disband. The advantages of the virtual chain include: low overhead, low risks, flexibility, and improved overall supply performance.

Sub-tier supply problems will never disappear. Potential problems with a small supplier far removed from a large multinational or end-product manufacturer can still endanger a multinational company through negative comments through social media (RBPS).

THE ORCHESTRA AS SUPPLY EXCELLENCE METAPHOR

The orchestra as well as an athletic team is often used as a metaphor for SCRM leadership and team excellence. The orchestra is a group of individuals, each of whom has a unique talent and a distinct way of delivering customer satisfaction. Usually, each player is a superstar in his or her right. Each musician as an individual and an orchestra player adds value and pleasure to the theater goer.

The parallels to supply chain excellence are strong. Each orchestra musician or supplier is chosen because of unique abilities and talents. The musician or supplier must pass an audition based on talent and ability. Once the musician or supplier is chosen, the musician must practice, rehearse, and continue to build upon his or her abilities, much like a supply-partner.

The orchestra is composed of different instruments and sections. This is similar to any supply chain with its functional groups, professionals, design processes, and product teams. The goal of each musical section, whether it is woodwinds, strings, or brass is to be consistent but also blend with other musical sections. Each professional may also add his or her own interpretation to the music much as a supply-partner may develop core capabilities and provide entire systems or assemblies.

The orchestra conductor, movie director, or operations manager is more often a team leader and coach than an authoritarian manager. The conductor or operations manager leads a group of professionals who are proficient with their instruments and have core abilities. The role of the modern conductor is to interpret the musical score and to shape the orchestra's sound so it pleases the audience. In the supply chain, the operations manager fine-tunes the supply stream so the final-customer is satisfied with the delivered product or service. The operations manager, much like the conductor, leads by interpretation, example, and strength of personality rather than by edict.[109]

INTERNAL CROSS-FUNCTIONAL COLLABORATION

Key process owners are responsible for their work output. The operations manager is sensitive to these requirements. The operations manager may have different SCRM objectives than the line manager or core process

owner. The line manager is under daily pressure to maintain process stability, meet quality requirements, satisfy dimensional specifications, keep
production up, improve process performance, ensure high-quality products, and meet regulatory requirements. The line manager is not responsible for upstream or downstream production. The result is that SCRM priorities may fall by the way side. So, a major responsibility of the operations
manager is to monitor overall SCRM process flows while being sensitive to
operational realities.

SUPPLIER OPERATIONAL CHALLENGES

While we have been addressing the end-product manufacturer challenges
the same occurs with supplier operations management. Supply operational managers must deal on a daily basis with the following:

- Conflicting end-product manufacturer requirements and messages.

- Demanding and irritable end-product manufacturers.

- Personnel turnover.

- Internal personnel issues.

- Grievances and other personnel challenges.

- Changing regulatory requirements and constraints.

- Safety and environmental compliance issues.

- Lack of resources.

- Spikes or troughs in end-product manufacturer demand.

- Internal process breakdowns.

The supplier also may have limited or no risk resources. The supplier may
have a small full time and a part time staff. The part time staff consists of
borrowed experts from production planning, manufacturing, quality engineering, or design engineering. Supplier personnel may report dotted line
to production management but are direct reports to their functional department or area.

CONTEXT: SCRM Matrix Organization Challenges

- Communication and coordination conflicts.
- No or little administrative support.
- Conflicts with operating groups.
- Multiple reporting paths.
- Multiple work flow paths.
- Possibility of changing priorities.
- Simultaneous and conflicting projects.
- Possibility of risk-control and power conflicts.
- High overhead due to parallel costs.
- Additional organizational layers.
- Group think.
- Excessive management.
- RBDM by consensus.
- Multiple and maybe conflicting policies and procedures.

Quality may provide a person to assist in supplier auditing, certification, and product testing. Human resources function may provide training or team facilitators for special SCRM projects. Production may provide first line supervisors to assist in conducting Failure Mode Effects Analysis, Statistical Process Control, capability, or other RBPS studies. Engineering may provide experts who develop technical specifications.

OPERATIONAL RISK MANAGEMENT

Every operations manager knows the expression 'on time, on budget, and within scope'. These are mission critical to all SCRM, sourcing, engineering, quality, materials, and logistics professionals. Well, these three issues: on time, on budget, and within scope are all risk factors relating to schedule, cost, and customer variances. Variance from target has evolved into another expression for risk. Cost, quality and schedule are the primary legs of supplier management. The conclusion is simple. Operations managers are also project risk managers controlling contract scope, cost, schedule (delivery), technology and quality variances.

The operations manager is evolving into a project risk manager. The product or service being sourced has a contract, product, or project lifecycle.

The project lifecycle starts with identifying final-customer needs, developing a project risk definition, and ending with the termination of the contract. It starts over again with a new product or service to be sourced. And throughout the contract, product or project life cycle, the operations manager addresses new contract terms, contract changes, and manages communications.

SCRM COMES DOWN TO PEOPLE

Experience is a hard teacher because she gives the test first, the lesson afterwards.
Vernon Law, Professional Baseball Player

As the velocity of technology increases, supply management organizations must adapt and anticipate these changes. The essence of an end-product manufacturer and its supply chain is constantly challenged. End-product manufacturers, organizations, and institutions have a body of culture, values, principles, and architecture that seem to value the status quo in direct challenge to the acceleration and disruption of technology.

MOVING TO THE CORE

Frankly, the Charles Handy work model discussed previously is an organizational challenge to find, retain and motivate the best operations managers. As Handy said:

> "The old model of the corporation was a piece of property, a piece of real estate. It was, quite simply, the property of its owners. ... The new model is based on an understanding that to hold people inside the corporation, we cannot really talk about them being employees anymore. To hold people, there has to be some continuity and sense of belonging. We also have to talk about two-way commitment – corporation to member, member to corporation"[110]

The Handy work model is devilishly simple. It provides the rationale to outsource non-core activities. When Handy first proposed his vision of the corporate workplace, it was considered too bleak and radical. Middle managers were considered indispensable. Many professionals and functions were considered necessary to sustain the organization. This has now all

---

CONTEXT: Rules for Leading the SCRM Transformation

- Establish the SCRM 'Tone at the Top.'
- Develop and follow a strategic SCRM vision.
- Clarify internal core competencies to architect, design, deploy, and assure SCRM.
- Conduct a SCRM readiness assessment.
- Get suppliers and other supply chain stakeholders engaged with SCRM.
- Lead and inspire the SCRM initiative.
- Communicate with critical stakeholders SCRM plans and objectives..
- Have a bottom-line, financial focus.
- Provide SCRM, RBT, RBPS, and RBDM training.
- Reward SCRM successes.

---

changed. All organizations are asking: "What is core and what is not." Core functions are retained while non-core activities are outsourced.

While purchasing and supply management are sometimes considered middle ring work, SCRM roles, authorities, responsibilities are more often migrating from the middle ring to the inner core. More end-product manufacturers recognize their supply chains are an untapped opportunity to impact the bottom-line and gain competitive advantage.

The challenge for many end-product manufacturers is finding supply executives with the right skill set to architect, design, deploy, and assure risk-controls. SCRM requires knowledge, skills, and abilities breadth and depth. As well, end-product manufacturers are looking for professionals with different knowledge skills and abilities as well as new perspective and lens by which to view at risk. The diverse skill sets and different lens allow for new ways to solve problems (RBPS) and make better decisions (RBDM).

SUPPLY CHAIN RISK MIDDLE MANAGEMENT
What happens to middle supply management, buyers, and agents who cannot evolve into the SCRM business ethic? As mentioned, middle operations managers assume different roles and have more accountability in

SCRM. If executive management considers middle level supply managers or buyers frozen and unresponsive, then obsolescent managers are right-sized or whatever the current euphemism is for being fired. If middle operations managers can become flexible, then they prosper with more responsibility and authority.

What happens to first-level, supply management supervision, the group of managers who are responsible for directing day-to-day supply operations? With the rise of employee involvement through self-managed, high-performance SCRM teams, the first line supply supervisor has a new role. The supply supervisor supports problem solving (RBPS) and decision making (RBDM) teams in his or her commodity or service area. The supervisor may become coach, facilitator, trainer, problem solver, assistant, motivator, or assume other roles. First-level supervision can make or break the SCRM initiative (RBDM).

FINDING AND RETAINING GREAT OPERATIONS MANAGERS

Developing organizational core competencies more often involves finding, nurturing, and rewarding the 'best and brightest' employees to evolve into supply chain risk managers. These are the organization's entrepreneurs, innovators, and dreamers. They are the people who create organizational and market value.

The 'best and brightest' operations managers are the means for deploying SCRM processes. In today's talent wars, end-product manufacturers are competing to find and retain these people. These high potential operations managers are induced through high salaries and generous options. The goal is not corporate paternalism but entrepreneurism where each operations manager has the opportunity to reach his or her potential and be accordingly rewarded. The hoped-for results include high retention, positive morale, exceptional loyalty, and intelligent risk taking.

The *Economist Magazine* distilled the importance of securing 'the best and brightest' operations managers:

> "With the life-cycle of products shrinking and competition coming from unexpected corners of the globe, companies have to be more

nimble than ever. This uncertainty helps some workers: if a firm has no idea from which direction the next competitive threat will come, one of its few sensible strategies is to amass good people to prepare for as many contingencies as possible."[111]

SUPPLY CHAIN RISK MANAGEMENT TEAMS

Very small groups of highly skilled generalists show a remarkable propensity to succeed.
Ramchandran Jaikumar

Operations managers have to learn new skills to be successful in the 'supply risk chain of the future' where integrated factories and suppliers instantaneously communicate to risk-control processes and monitor products.

We believe that within a few years, more highly trained, technical workers will fill many SCRM positions. Why? The consequences of supply chain failure are so high that the profession will require a diverse and technology friendly group of SCRM professionals. If supply chain or plant managers need information, they will tap directly into central computers without any need of middle managers to collect, integrate, and communicate the information. As well, artificial intelligence (AI) and machine learning will allow operations professionals and operations managers to have instant access to supply chain and supplier information.

ANATOMY OF SCRM TEAMS

SCRM teams come in many shapes and forms. Teams are loose clusters of individuals or they are highly structured, customer-supply project teams bound by contracts and mutual obligations. They are self-managed or are tightly controlled. They are permanent or temporary. A team's shape is flexible depending on risk skills, purpose, SCRM objectives, culture, company politics, technology, and other factors.

Quinn Mills in **The Rebirth of the Corporation** describes the new organization as clusters of teams that share the following:

- Handle their own administrative functions.

- Develop their own expertise.

- Express a strong final-customer orientation.

- Are project and action oriented.

- Share information broadly.

- Accept accountability for business results.[112]

USE OF MULTI-SKILLED SCRM TEAMS

There are many types of SCRM project teams. An executive project team may develop a SCRM strategy. Divisional or business unit teams may negotiate a partnering arrangement with suppliers, develop customer-supply selection procedures, deploy elements of a supply chain risk initiative, or develop risk procedures to improve supplier risk quality. Plant level teams may pursue ISO 9001:2015 certification, monitor incoming products, deploy ISO Risk Based Thinking (RBT), or integrate a new machine into the production line.

Entire supply chains may be structured in self-managed, SCRM teams. To realize the potential of technology, end-product manufacturers are integrating supply chain stakeholders into these multidisciplinary teams. In these systems, semiautonomous teams of five to seven multidisciplinary stakeholders work on issues spread across the chain. Teamwork, participative managers, and multiskilled workers are keys to success. The payoff is SCRM designed around participative and innovative teams is 30% to 50% more productive than conventional methods of organizing work.[113]

Self-managed SCRM teams share common goals and accept responsibility for production schedules, risk management, quality, cost control, upgrading professional skills and assignment of work. Previously, purchasing may have been responsible for these activities. Now, multidisciplinary SCRM teams direct these activities.

END OF THE LONE PROFESSIONAL

The lone professional is dead in all organizations - the lone person developing killer software, the lone salesperson selling million dollar systems, or the lone supply management officer visioning the future of a billion-dollar supply chain. No one person has the skills to inspire and know all things.

There is too much diversity and access to resources and information. No one person is as smart as a motivated SCRM team.

An example may help to illustrate this point. In the supply chain, work is arranged around end-to-end processes, rather than specific tasks. A semi-autonomous SCRM team of five to seven multiskilled stakeholders may manage a process or manage risks that span across several functional departments and go into the supply stream. Functional specialists from engineering, ERM, sourcing, or manufacturing engineering may participate or support this team on an as-needed basis. RBPS and RBDM will require broad based jobs, teamwork, participative managers, and multiskilled workers.

CHIEF SUPPLY MANAGEMENT OFFICER (CSMO)

Great leaders are almost always great simplifiers who can cut through the argument debate and doubt to offer a solution everybody can understand.
Colin Powell

The premise of this book is that purchasing and supply management are evolving into SCRM . The head purchasing position used to be a manager and sometimes a director. More often, end-product manufacturers are moving supply management to an executive level, supply chain or Chief Supply Management Officer (CSMO).

FROM DIRECTOR TO VICE PRESIDENT
We are now seeing VP's and above assuming the title of Chief Sourcing Officer or Chief Supply Management Officer. Titles are important because they reveal the importance an end-product manufacturer places on the function. The new CSMO is an Executive Vice President, Senior Vice President or Vice President.

The CSMO plays the mission critical role of providing 'make or buy' input, setting supply chain risk governance expectations, shaping the supply chain business model, developing the SCRM strategic vision/plan, communicating SCRM inducements, offering supply innovation assistance, and developing the supply base.

CSMO VALUE PROPOSITION

One of the first challenges of all end-product manufacturers is the 'make or buy' decision. The CSMO knows supply base capabilities and its ability to add strategic value. The CSMO is brought into strategic discussions determining core competencies and what should be outsourced or insourced. Once a sourcing decision is made, the CSMO provides the political will to make the hard and fast policies to whom, why, what, how, and when products will be outsourced (RBDM).

This is not happening easily. I have heard a lot of grumbles from old line purchasing agents, buyers, and supply managers who say these new folks have not paid their dues. I hear the following from traditional purchasing agents: "Do these new sourcing people have the savvy to sniff out supplier promises and make fundamental business decisions on expensive purchases?" These new people probably do not have the negotiating skills, relationships, and business skills of an old-time sourcing pro. But, they bring new skills to the table such as evaluating the effectiveness of a supplier's Six Sigma program, constructing intelligent supply AI, deploying SCRM systems, designing a lean manufacturing initiative, and integrating the supplier's best design practices into the end-product manufacturer's product development (RBDM).

So, what will the new SCRM function look like? This is evolving. However, we can make some early observations. More often, SCRM is at the center of the supply chain, the point at which it is managed and risk-controlled. But, this is only part of the supply chain equation. Supply chain stakeholders also include production control, quality engineering, risk management, IT, design engineering, logistics, distribution, production scheduling, inventory management, and demand forecasting.

INCREMENTAL CHANGE AND TRANSFORMATION

A critical SCRM question arises: Should the SCRM journey be incremental or a transformative as shown in the above Figure 30 . There is no easy answer, but depends on the context of the disruption and risks facing the end-product manufacturer.

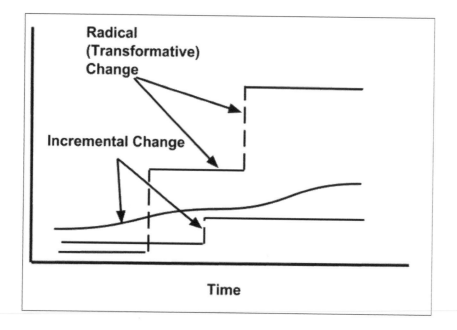

*Figure 30: **Transformation Vs. Incremental Change***

Organizational change and transformation processes are different. According to Richard Pascale, author of **Managing on the Edge**, change is 'incremental improvement' while transformation is a 'discontinuous shift in capability.' He defines transformation as a discontinuous shift in bottom-line results, industry standards, benchmarks, and employee perceptions. From this definition, SCRM by some is seen as a discontinuous shift from transactional purchasing to enterprise SCRM.

The transformation process can result in major supply chain changes that are brought about by engaged stakeholders. These stakeholders view transformed customer-supply relationships and expectations as a different SCRM business model from even 5 years ago. While incremental change is sufficient for many well performing companies, a supply chain in a highly competitive market segment may have to transform continuously.[114]

CHANGE MANAGEMENT
How to manage in VUCA time is a core question for all end-product manu-facturers. We believe that general management is moving to RBPS and

RBDM. So, what does this really mean? Each element of VUCA, specifically volatility, uncertainty, complexity, and ambiguity is altering what we mean by management. Take a look again at the 'Old School' to 'New School' management sidebars in Chapter 2. They reveal some of the changes facing end-product manufacturers.

Managing supply change is a difficult process. The process has to respect supplier rights, comply with regulations, comply with work agreements, be ethical, and in the end be effective, economic, and efficient. Supply chain change becomes more difficult as international boundaries, financial exigencies, political constraints, cultural pressures, and other factors increase.

Disruptive transformation and SCRM change may involve a certain amount of organizational, management, supplier, and individual discomfort. The challenge is to institutionalize discomfort so it becomes part of the changing supply chain fabric, ethic, and culture. In other words, marketplace pressures and end-product manufacturer requirements are communicated throughout the supply chain so stakeholders become flexible and anticipate the need for change.

FEAR FACTOR PARALYSIS
This is difficult for end-product manufacturers because extreme, external pressures can stymie initiative. End-product manufacturers become frozen through fear of potential downsizing. Suppliers become frozen through fear of being eliminated. The supply chain is redesigned.

Organizational resistance is normal and expected when the supply chain is formed and end-product manufacturers must change. Resistance may occur initially, disappear, or continue throughout the supply change process. What is the supply change manager or team supposed to do with pockets of resistance? Most supply chain stakeholders are rational and understand the need for change if it is plainly explained and if the urgency is particularly evident. Change may also be induced through financial incentives, training, promotion, or other mechanisms.

SCRM is not for the faint-hearted and end-product manufacturers may not want to and simply cannot change. Kobe Steel was one of Japan's preeminent steel companies. However, it faked its quality guarantees and certificates for more than 10 years because of competitive pressures. This shocked Japan's quality culture. Media throughout the world chastised Japan's vaunted national quality culture. The UK's *Telegraph* newspaper had headlines that shouted 'Kobe Steel Quality Scandal Driven by Pursuit of Profits and Demanding Corporate Culture.'[115]

LEADING THE SCRM CHANGE

A challenge to all executives, including the chief supply executive, is managing change. For some managers, this involves being proactive, embracing change, searching for new opportunities, and capitalizing on them. For others, this means surviving the latest management fad or simply maintaining purchasing head count in a downsizing workplace.

The essential element behind any supply chain transformation is a person or team that catalyzes the SCRM transformation. Often, the CSMO works with a core group of senior executives to deploy the SCRM initiative. If executive management and the Board of Directors make the SCRM transformation a priority, devote time, assign resources, and manage it then it has a much higher chance of success.

SUCCESSFUL SCRM TRANSFORMATIONS

Successful SCRM transformations usually require:

- Visceral understanding by the Board and executive management of the consequences of VUCA disruption and the need for SCRM design, deployment, and assurance.

- Board reporting of operations management and supply chain risks.

- Sense of competitive urgency to manage supply chain risk and to use it as a competitive differentiator.

- Sense of business and supply chain risk alignment.

- Active CSMO and executive management involvement.

- Long-term SCRM view and mandate.

- Involvement and acceptance by most or all organizational stake-holders.

- Extensive individual risk training and supply development.

- Extensive efforts and resources for SCRM design, deployment, risk-controls, and risk-assurance.

SCRM TRANSFORMATION PROCESS

The SCRM transformation first starts internally by 1. Establishing supply chain risk drivers; 2. Defining clear SCRM objectives; and 3. Working to eliminate supply chain inhibitors and risks. Supply chain transformation drivers include executive management commitment, final-customer satisfaction, and competitive focus. SCRM objectives are defined at the: 1. Enterprise level; 2. Programmatic/Project/Process level; and 3. Transactional/Product level. Supply chain inhibitors or risks are elements that can obstruct the SCRM initiative. Inhibitors may include departmental resistance, internal functional boundaries, supplier fears, politics, and other forms of resistance and risks.

Depending on organizational culture, context, business model, and urgency, the transformation can be collaborative or even some form of 'command and control.' We have seen and experienced both. We prefer the collaborative approach. However, more end-product manufacturers believe they should disrupt preemptively, rather than wait for the organization to be disrupted by external pressures.

If executive management feels comfortable with command and control, the driver approach is first used internally. Executive management will establish the SCRM mandate for change and then start integrating various SCRM elements including supply development, ERM, RBT, RBPS, RBDM, Six Sigma, Enterprise Resource Planning, lean, or ISO 9001:2015 certification throughout the organization and then into the supply stream.

This is not an easy process. Driving a SCRM process can cause additional organizational stress and strain. It is more effective to use a combination

---

CONTEXT: SCRM Executive Characteristics

- SCRM governance knowledgeable.
- ERM and SCRM architecture, design, deployment, and assurance knowledgeable.
- Flexible and adaptable.
- Excellent communicating abilities.
- SCRM leadership by example.
- Persistent.
- Persuasive.
- Politically savvy.
- Relationship building abilities.
- Excellent planning and end-product manufacturer skills.
- Can do attitude.
- Cooperative.

---

of collaborative drivers and inducements to shape and direct the SCRM transformation.

Freeing the end-product manufacturer from existing ways of conducting business is difficult but is recognized as an effective means for inducing change. In general, it is wiser to proceed slowly, show daily commitment to the change process, understand actions speak louder than words, couple SCRM with empowerment, and reinforce actions that support the SCRM initiative.

SUPPLY CHAIN DISRUPTION

Supply base disruption or even reduction is now a fact of business life and is an integral element of the SCRM initiative. This creates new and worrisome challenges for operation managers and suppliers as they try to figure out new SCRM rules. Unfortunately, consequencers and negative reinforcers seem to be the common means for forcing or inducing supply change. However, these breed fear. Supply management priorities become unclear during such times. Everyone is looking for cost cutting opportunities, which may conflict with the overall customer-supply partnering messages and opportunities.

What is the role of SCRM professionals in these turbulent times especially if he or she is the change manager? The role of the supply change manager is difficult in the best of times and is very difficult in transformational and disruptive times. In change-resistant end-product manufacturers, the operations manager may feel she or he has a bulls-eye on her back. Why? Change is transformational instead of evolutionary. Executive management may not totally understand and support the SCRM initiative. There are pockets of resistance. The end-product manufacturer is not adaptable or people do not have the skills to affect change.

Cynicism becomes widespread over the newest fad. People already are too busy on day-to-day operations and do not have time for the new SCRM project or methodology. The wrong person may have been chosen to lead the supply change initiative to risk. Whatever the cause, the result is the same. The SCRM transformation and initiative tend to bog down and get sidetracked. Key people including SCRM professionals become discouraged over the floundering initiative.

SCRM LEADERSHIP ROLES

Leadership is action, not position.
Donald McGannon

I thought about this for a long time. 'Succeed or perish' explains the prevalent thinking of many senior executives. This applies to the end-product manufacturer as well as to the executive management including CSMOs. Think about it? There are no stupid, long-term senior executives. Most perish quickly upon making several bad decisions. Hence, there is more focus on RBPS and RBDM.

SCRM GOVERNANCE

An end-product manufacturer has many working components and stakeholders with differing requirements. In architecting SCRM, it is critical to know who will benefit from SCRM and RBT and develop the business case to these requirements. As well, understand concepts such as Governance, Risk, and Compliance (GRC) and Enterprise Risk Management (ERM) because they will become integral to the SCRM initiative. The business case

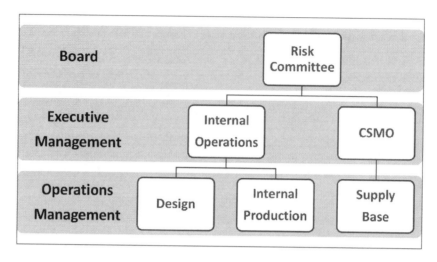

Figure 31: SCRM Deployment

provides the rationale for SCRM in terms of costs, monetary benefits, success factors, and Key Risk Indicators. The business case provides the financial justification to architect, design, deploy, and assure SCRM.

Executive management is concerned about identifying, managing operational risks, and reporting the successful mitigation of risk. Executive management and the Board of Directors want to assure investors and other stakeholders that they are focused on profitability, know how to achieve profitability, understand how to remove impediments (risks) to profitability, and develop appropriate risk-controls to sustain profitability. Secondly, the end-product manufacturer wants its stakeholders and interested parties to know that it embraces good governance, SCRM, RBT, ERM, and compliance with applicable statutes as illustrated in Figure 31.

We have identified more than 17 sectors moving to ERM, which involves some form of SCRM. ERM statutes often place the responsibility of managing supplier risk clearly at the Board level, such as the U.S. Federal Reserve System:

> "The use of service providers does not relieve a financial institution's Board of Directors and Executive management of their responsibility to ensure that outsourced activities are conducted in a safe-and-sound manner and in compliance with applicable laws

and regulations. Policies governing the use of service providers should be established and approved by the Board of Directors, or an executive committee of the Board. These policies should establish a service provider risk management program that addresses risk assessments and due diligence, standards for contract provisions and considerations, ongoing monitoring of service providers, and business continuity and contingency planning."[116]

CSMO ATTRIBUTES
What makes a CSMO leader? Is a leader made, nurtured, or self-selected? Leadership is still difficult to understand and to explain. Increasingly, leaders are people who guide themselves or a group to do what needs to be done as well as reach ever-higher goals. In general, CSMOs are normal people who possess high energy, are committed, can share responsibility, have high-values, are technical, and are highly credible.

The CSMO is a leader who can learn and adapt to changing VUCA circumstances. Another element of leadership is the ability to communicate at a visceral level. The CSMO does not simply communicate supply information but has the ability to communicate the SCRM vision. Often, this communication means being actively involved in strategic and tactical discussions. The CSMO questions, listens actively, and surfaces issues that are important to individuals, to supplier teams, and to the end-product manufacturer.

An interesting question is whether SCRM leadership can be learned and if it can be shared. Pursuing and sharing leadership is a critical element of successful SCRM team problem solving (RBPS) and process innovation. SCRM teams have the responsibility and authority to ensure the supply chain runs smoothly. This message is deployed throughout the chain so critical stakeholders can improve their sub-tier chain processes.

FLATTER ORGANIZATIONS
Successful SCRM requires that operations managers have the authority and responsibility to take risks and make intelligent decisions. Buyers are elevated to operations managers with more decision making (RBDM) authority.

Rigid, hierarchal, and authoritarian styles of supply management are becoming history. Operations managers are encouraged and rewarded to make decisions as part of a SCRM team. The SCRM organization is more fluid, some say virtual, as SCRM teams are established to get a job done and are then disbanded (RBDM).

SCRM ROLES AND RESPONSIBILITIES

The company with the second-best organization ends up second place in the market.
D. Wayne Calloway, CEO PepsiCo, Inc.

SCRM requires new leadership and management abilities. Let us look at some of the new roles and responsibilities SCRM requires:

- Risk governance and compliance officer.

- SCRM leader.

- SCRM architect.

- SCRM role model.

- SCRM champion.

- SCRM initiative and project sponsor.

- Change manager.

- Ombudsperson.

- Manager.

- Risk coach.

- Risk organizer.

- Enabler.

- Risk cheerleader.

- Entrepreneur.

- Translator.

- Engineer.

- Internal risk consultant.

RISK GOVERNANCE AND COMPLIANCE OFFICER

In the U.S., SCRM and ERM are moving into statutes and rules. U.S. federal and state risk regulations impact almost every element of the supply chain involving packaging, safety, health, and so on.

For example, most if not all auto end-product manufacturers are moving to electric and then to autonomous vehicles (self-driving cars). End-product manufacturers are working with their supply base to facilitate the transition from gas and diesel cars. This is a Board level and executive challenge. Daimler (AG) issued the following warning in a recent annual report on the risks the transition may incur:

> Due to the planned electrification of new model series and a shift in customer demand from diesel to gasoline engines, the Mercedes-Benz Cars segment in particular is faced with the risk that Daimler will require changed volumes of components from suppliers. ... This could result in over- or under-utilization of production capacities for certain suppliers. If suppliers cannot cover their fixed costs, there is the risk that suppliers could demand compensation payments. ... Necessary capacity expansion at suppliers' plants could also require cost-effective participation." [117]

SCRM LEADER

The CSMO understands end-product manufacturer business model requirements and knows how to configure the supply base to satisfy these requirements. The CSMO or senior operations executive must integrate supply base core competencies with those of the end-product manufacturer, direct the SCRM function, and develop 'world-class' suppliers to meet these requirements. The CSMO may have to redesign the SCRM function and hire new operations managers who can develop, organize, motivate, coach, train, mentor, and energize suppliers towards SCRM.

SCRM leadership and supply management may seem synonymous. However, SCRM leadership is an art. Depending on the requirements of the supply chain, market, organizational culture, and abilities of the people, leadership may involve 'command and control' or 'coaching and mentoring.' More often, SCRM leaders are closely involved with suppliers and employees, suggesting, and demonstrating as opposed to directing what needs to be done.

SCRM managers may lead through example rather than through hierarchal or positional authority. As responsibilities and authorities have been downloaded to people doing the work, SCRM functions have been downloaded to operating units. The operations manager may not be able to select a supplier or dictate a supply innovation initiative. Enlisting the support of supply chain stakeholders requires new management skills. The operations manager may have to lead by example and support operating personnel across functional areas and work processes. The operations manager must develop relationships, trust, and credibility with supply management.

SCRM ARCHITECT

The operations manager may have to architect, design, deploy, and assure SCRM down the supply chain. The operations manager works with SCRM stakeholders to reengineer processes to control risks, achieve Key Risk Indicators, manage information flows, develop technical risk systems and other processes. The conventional wisdom is that end-product manufacturers must transform processes and streamline work flows to become globally competitive. Supply chain redesign is a never-ending process to accommodate new products or satisfy new end-product manufacturer requirements.

SCRM ROLE MODEL

Integrity as a moral principle is often cited as a guide for organizational governance and personal maturity. Integrity is also an essential element of all SCRM initiatives. Integrity is the ability to honor commitments with company stakeholders, suppliers, final-customers, employees, and other critical supply chain stakeholders. Integrity is the ethic that reinforces an organization's vision, mission, and values. The lack of integrity is often the reason SCRM initiatives have failed while others have flourished.[118]

SCRM CHAMPION

The SCRM champion is the CSMO, team of senior executives, or even the Chief Executive Officer. In formal and informal settings, the champion reminds the organization of the importance of SCRM. The champion provides active and consistent support for SCRM activities.

An operations manager must be a mentor and be capable of being mentored. For example, mentoring suppliers is fundamental to SCRM development and developing long-term supply relationships.

SCRM INITIATIVE AND PROJECT SPONSOR

A consistent mantra in this book is that a senior executive should sponsor the SCRM initiative. In a large end-product manufacturer, the SCRM sponsor is the CSMO who guides supply operations managers. If SCRM is new to the end-product manufacturer then the executive sponsor can:

- Provide access to the Board and executive management.

- Provide SCRM resources.

- Handle major supply risk problems (RBPS).

- Provide instant credibility to the SCRM project and supply development initiative.

- Keep executive management informed of supply changes.

- Sit in on senior level meetings.

- Vocally and actively support the SCRM initiative.

CHANGE MANAGER

Change manager is a relatively new term to describe the role of a person or a group to move the end-product manufacturer in a new direction. The CSMO is the enterprise level leader, while operations managers are the change managers for their commodity suppliers. Sometimes, the initial change manager is the CEO who assigns the supply transformational responsibilities to the CSMO.

Usually VUCA or a competitor signals the need for change. Old ways of business are no longer sufficient to be competitive. For example, an end-product manufacturer may change from an engineering focused to a customer-focused supply chain organization. Or, it may have to evolve from a product, transactional orientation to a SCRM enterprise orientation.

OMBUDSPERSON

In some end-product manufacturers, the operations manager serves as the ombudsperson on tactical customer-supply risk issues. In much the same way, the human resources manager is the ombudsperson on organizational and cultural matters. The operations manager, as an ombudsperson, is the designated person to resolve risk-control, quality, customer satisfaction, delivery, and customer-supply issues among teams, departments, and business units.

MANAGER

An operations manager has traditional managerial functions, specifically to:

- **Plan.** Operations manager communicates commodity, product, technical, and risk requirements to suppliers.

- **Organize.** Operations manager organizes the flow of risk-control information and requirements to the supply base.

- **Risk-control.** Operations manager establishes, monitors, and intervenes if required to risk-control, supply processes and if necessary correct them.

- **Directs.** Operations manager directs and oversees SCRM team risk activities.

- **Staff.** Operations manager hires, develops, and evaluates supply staff performance against Key Risk Indicators.

RISK COACH

The relationship between SCRM and sports management is very close. End-product manufacturers are evolving into groups of self-managed SCRM teams. Team structures are flat. The operations manager is more

often a coach. The operations manager works through his or her employees. Coaching requires the ability to encourage teamwork, motivate employees, and establish SCRM tactical direction.

In the athletic metaphor, the coach is nominally in charge. The coach decides who is on the team, who plays, and so on. But the coach is not
responsible for the big play or the series of downs that scores points. The
coach broadly establishes the game plan, provides training, encourages
self-sacrifice, mentor's players, and serves other facilitating roles. Today's
operations manager may well be the coach of the SCRM team composed
of multidisciplinary experts.

Operations managers and coaches have similar goals to be competitive
and to win games. This requires getting the best from players. Success
depends on how players follow through on supply management assign-
ments.

RISK ORGANIZER

A major responsibility of the operations manager is to interface, organize,
and monitor supply development and innovation projects. The operations
manager is responsible for organizing, facilitating, monitoring and reporting
project progress. The operations manager also interfaces with operational
units such as engineering, manufacturing, planning, and other functions to
organize SCRM activities.

ENABLER

The operations manager, as enabler, assists functional personnel and process owners to establish and meet SCRM objectives The enabler provides
assistance and counsel to process owners and SCRM stakeholders. This
may mean developing supply risk-control plans, evaluating product plans;
monitoring supply risk-control processes; training supply personnel; monitoring risk-controls and deployment; and supporting SCRM initiatives. The
SCRM enabler walks a fine line. The enabler can provide assistance, risk-
assurance, and guidance but cannot take responsibility away from process
owners.

RISK CHEERLEADER

Cheerleading managers are optimistically realistic. Cheerleading may involve several activities such as uplifting SCRM talks, encouraging supplier risk teams, recognizing supplier risk accomplishments, thanking personnel, being a risk mentor, or defusing customer-supply risk tensions. They are infectious self-starters who generate enthusiasm in others. Cheerleader management is not rose colored or Pollyannaish. Cheerleader management is tightly focused on raising awareness and pursuing SCRM innovation.

Supply chain cheerleaders focus energies on meeting the end-product manufacturer's business risk mission and risk-control objectives. They empower SCRM teams to target their energy and the company's resources on opportunities for success and high promise outcomes.

Cheerleading is one of the more difficult challenges for the operations manager. Disillusionment over a SCRM initiative can take its toll. Cynicism or defeatism can result and become self-fulfilling. The operations manager must cheerlead when a project flags because of rising expectations, poor results, internal resistance, conflicting projects, ego-gratification, selfishness, poor communications, risk-control ineffectiveness, or slow SCRM progress.

ENTREPRENEUR

Several years ago, there was a trend to outsource supply management functions. Internal users became customers. Supply management departments became service departments. They had to become customer, cost, and value sensitive. The supply management function became a small business and the operations manager became an entrepreneur. Supply management billed user departments for supply training, purchasing, consulting, management, and other services. These services had to add value and be competitively priced. Purchasing had to compete directly with external providers of the same services. This ensured internal staff provided value. If operating groups did not want to use the supply department's services, the message was loud and clear - no business, no monies, no purchasing personnel.

TRANSLATOR

Each discipline has its own language including mechanical engineering, supply management, finance, and sales. The language of SCRM is a process based driven language involving Enterprise Risk Management, lean, Six Sigma, just-in-time, ERP and other supply concepts.

ENGINEER

Operations managers are more often pulled into new product development since many of the parts in the new product will be sourced and SCRM is a critical element of sourcing. Since operations managers must find the most capable supply-partners to bring into product development. Operations managers of high-tech products often have a technical background. This helps operations managers who must source long-lead technical products and components.

INTERNAL RISK CONSULTANT

The SCRM function is often a staff function. The function may have few full-time employees and a small budget. The department must leverage itself through operating departments to get the job done. The supply management department may not have the staff to fully lead and support major SCRM development initiatives, such as RBPS, RBDM, ISO 9001:2015 lean, Six Sigma quality, ERM, and customer-supply risk partnering.

As more supply management responsibilities are operationalized, what is the role of the operations manager regarding SCRM? The operations manager may become an internal consultant. The manager works with operational groups, suppliers, process owners, and plant managers to initiate SCRM initiatives and install risk-control mechanisms to ensure a smooth flow of products. This role requires a special individual who has 'real world' consulting expertise, who can discover new opportunities for SCRM.

SUMMARY

Operations managers must live in a constantly changing VUCA world. Why? The marketplace is global. End-product manufacturer requirements are changing in the U.S. and throughout the world. Suppliers in developing nations are becoming formidable competitors with their own supply chains. There are no uniform rules for SCRM success.

Transformations, by their nature, are not easy. An end-product manufacturer or supplier going through a SCRM transformation, not an incremental change, may experience a wrenching and abrupt process.

Siloed problem solving (RBPS) and decision making (RBDM) presents another risk in SCRM. An end-product manufacturer will assume that engineering is responsible for product design. Quality is responsible for the quality of incoming products. Supply management is responsible for securing appropriate products just-in-time to be used. RBPS and RBDM integration become essential.

Appropriate stakeholders are all involved in making decisions regarding supply chain risks. Silos are broken down. White space risks between functions are clarified. Process of making decisions and solving problems are clear. Problem solving (RBPS) and decision making (RBDM) assumptions, inputs, process, and outputs are clear as well.

The basic theme of this chapter is that SCRM processes are improved over time. Operation managers must know basic process and risk auditing tools; speak the language of supply risk-control development; and know internal/supplier process risk capabilities. As well, operations managers are sensitive to different end-product manufacturer cultures. Operations managers also assume different roles during supply development.

NEXT CHAPTER
In the next chapter, we discuss how to projectize SCRM deployment.

CHAPTER 9:
SCRM DEPLOYMENT -
PROJECTIZING

WHAT IS THE KEY IDEA IN THIS CHAPTER?

Deloitte conducted a study of 600 C-level executives that revealed 45% felt their SCRM programs are only somewhat effective or not effective at all. A mere 33% even use risk management approaches to strategically manage supply chain risk.[119] And, supplier quality continues to be the #1 supply chain risk according to many surveys.

The key idea in this chapter is that SCRM deployment can follow a logical projectizd process, which is described in this chapter.

SCRM PROJECT DEPLOYMENT

Businesses aren't run by geniuses. It is a matter of putting one foot after another in a logical fashion. The trick is in knowing what direction you want to go.
James R. Barker

There is no one officially sanctioned or approved SCRM deployment methodology. It seems that every consultant to differentiate himself or herself recommends something a little different. The following SCRM methodology as illustrated in Figure 32 is a plain vanilla model that incorporates steps that work:

- **Step 1.** Understand the competitive environment/context.

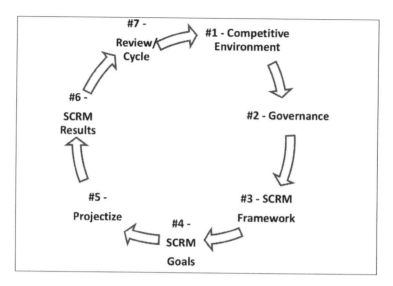

Figure 32: **SCRM Deployment Project Lifecycle**

- **Step 2**. Establish SCRM governance and risk management.

- **Step 3**. Develop internal SCRM processes.

- **Step 4**. Identify SCRM objectives, KPI's, and KRI's.

- **Step 5**. Deploy SCRM project results.

- **Step 6**. Evaluate SCRM project results .

- **Step 7**. Review and start again (Go back to Step 1).[120]

STEP 1: UNDERSTAND COMPETITIVE ENVIRONMENT/CONTEXT

Times change and we change.
Latin Proverb

A supply chain exists in a competitive, business, and cultural context. The purpose of competitive analysis is to identify competitor's supply chains and gather information about the environment in which a supply chain competes. Ultimately understanding the competitive environment facilitates the key 'make or buy' decision.

Competitive analysis is an ongoing process. As an end-product manufacturer develops and deploys SCRM strategies, the results are monitored on an ongoing basis to ensure the right things are done right at the right time. Continuous feedback monitoring allows for ongoing changes in make/buy decisions, supply development, and technology application (RBDM).

A critical question in the competitive analysis is how to acquire core SCRM competencies. One method is through training. Another is through supply chain alliances that reinforce internal core competencies and strengthen the business mix. If an end-product manufacturer acquires a business or a supply-partner, it expects a synergistic effect of related businesses to reinforce the core systems.

An end-product manufacturer must conduct a competitive analysis identifying what are its unique value-added differentiators of its competitors. For example, a consumer electronics company may focus on its higher-end products with higher margins rather than commodity retail products with lower margins. Or it may sell low-margins products and make profits on its higher end services such as Best Buy with its Geek Squad support services.

KEY ELEMENTS IN THIS STEP OF THE SCRM JOURNEY
SCRM factors to consider in this step include:

- Identify the major economic, social, cultural, and technical trends and risks impacting the business and supply chain over the next year and three years.

- Identify competitive companies and competitive characteristics of their supply chains.

- Identify their relative strengths and weaknesses.

- Locate competitor's points of weakness and possible areas with the greatest competitive leverage.

- Determine 'make or buy' and how and where to compete.

STEP 2: ESTABLISH SCRM GOVERNANCE AND RISK MANAGEMENT

Many of our best opportunities were created out of necessity.
Sam Walton, Founder of Walmart

SCRM is a total business, organizational, and technical approach towards developing internal and supplier systems, processes, and products that please final-customers. Executive management commitment and involvement are essential for SCRM to flourish. Executive management takes the lead for creating a flexible SCRM environment, defining the culture, and encouraging change.

The Board and executive management are responsible for:

- SCRM Vision.

- Long-term commitment of SCRM.

- People involvement and empowerment of SCRM process owners.

- SCRM goals and objective definition.

- SCRM methodology and risk management.

- SCRM training.

SCRM Vision

Executive management and critical stakeholders establish a vision of supply chain possibilities. The visioning process links external market requirements, opportunities, and challenges to end-product manufacturer core process capabilities. If internal capabilities do not exist, they are developed internally or acquired from a supplier.

LONG-TERM COMMITMENT OF SCRM

SCRM is a business model and requires a long-term commitment to ensure its success. End-product manufacturer requirements change over time and the SCRM model also changes to accommodate today's end-product man-ufacturers as well as tomorrow's. For example a few years ago, supply management was largely defined in terms of low-price and ability to meet ISO 9001 conformance requirements. Now, niche final-customers must be pleased with customized, cost competitive, aesthetically pleasing products and services that are delivered courteously and quickly.

Executive management commitment involves several visible activities, specifically providing:

- Long-term SCRM perspective and drive.

- Focus on SCRM innovation and continuously improving business results.

- SCRM policies defining the direction of the end-product manufac-turer and supply chain.

- Supply chain standards specifying what is expected and accepted.

- SCRM talks, speeches, and rewards.

- Monies, facilities, people, equipment, and other resources.

- Value-added SCRM training.

- Internal and supply recognition for exceeding end-product manu-facturer requirements.

PEOPLE INVOLVEMENT OF SCRM PROCESS OWNERS

People are the means of deploying SCRM. Internal professionals and sup-pliers own their processes. Executive management establishes, supports, and reinforces supply chain initiatives by rewarding stakeholders to submit ideas and to deploy them within the supply chain without fear of reprisal.

SCRM sometimes requires a change in how business is conducted if not an organizational and cultural transformation. It takes time and patience. The process may start with low-level supply participation and move toward more supply involvement, empowerment and finally customer-supply pro-cess integration.

The critical elements of success are faith, trust, and results. Faith is required to start the SCRM process. Trust between the end-product manufacturer and supplier is required to ensure the process will proceed satisfactorily. Finally, demonstrable results ensure improved SCRM capabilities.

SCRM GOALS AND OBJECTIVE DEFINITION
The goals and objectives of the SCRM initiative must link, support, and reinforce the overall SCRM vision, mission, plans, and objectives. Supplier linkages are definable, demonstrable, timely, and measurable.

Until recently, SCRM initiatives were approved because they were viewed as good for the end-product manufacturer. Monies were allocated and spent with sometimes few results. Now, executive management and other stakeholders are smarter and regard SCRM as an investment that must demonstrate immediate and sufficient financial return beyond the Business to Business (BTB) hype. Most importantly, these returns must support the critical enterprise mission.

SCRM METHODOLOGY AND RISK MANAGEMENT
SCRM must follow a disciplined and integrated approach using the appropriate sourcing tools, risk management methodologies, risk principles, and risk assessment techniques. The right approach and tools must consider the culture of the supply chain organization. SCRM professionals must monitor the right supply chain control variables, recognize when there is an abnormality, intervene with the right tools, and remove the deficiency. This is called management by exception.

SCRM TRAINING
Changing end-product manufacturer requirements and new product development require continuous knowledge management and stakeholder training. New suppliers are trained in how to integrate core capabilities with those of the end-product manufacturer's. Internal employees are trained in SCRM tools, techniques, culture, customer requirements, processes, systems, and products.

KEY ELEMENTS IN THIS STEP OF THE SCRM JOURNEY
SCRM factors to consider in this step include:

- Establish an end-product manufacturer governance, risk management, and compliance (GRC) structure that integrates SCRM requirements and incorporates these requirements into organizational policies.

- Establish a set of SCRM accountabilities and responsibilities that ensures stakeholders understand and use RBPS and RBDM.

- Develop SCRM Tone at the Top and risk culture based on organizational context.

- Obtain Board, executive level, and CSMO support.

- Review supply chain governance and risk appetite in light of the end-product manufacturer's risk-controls of critical suppliers.

- Review end-product manufacturer's current SCRM business model and assumptions.

- Develop SCRM vision, mission, and strategic plan.

- Define what SCRM means to the end-product manufacturer and the benefits accrued to the end-product manufacturer and suppliers.

- Establish an end-product manufacturer governance structure that integrates SCRM requirements and incorporates these requirements into the organizational policies.

- Assemble an interdisciplinary team that takes an enterprise-wide view of the supply control environment and supply risk-controls.

- Develop designed and tailored risk taxonomy and definitions for the end-product manufacturer.

- Develop Key Risk Indicators (KRI's) and Key Performance Indicators (KPI's) for the SCRM initiative.

- Understand customer-supply partnering requirements and expectations.

- Follow a consistent process of selecting, certifying, monitoring, and improving suppliers.

- Ensure all stakeholders and interested parties understand and can meet their SCRM objectives.

- Identify suppliers that provide critical products and services as initial SCRM partners.

- Build a cross-functional, SCRM team to manage the SCRM initiative.

STEP 3: DEVELOP INTERNAL SCRM PROCESSES

A relentless barrage of why's is the best way to prepare your mind to pierce the clouded veil of thinking caused by the status quo. Use it often.
Shigeo Shingo

The supply chain is a primary process with sub-processes. To understand the supply chain and its various tributaries, the chain should be process and decision mapped. Most readers already know how to conduct process and value stream maps of material flows. Decision process maps are relatively new.

DECISION PROCESS MAPS
Decision process mapping show what, where, and how critical problems are solved (RBPS) and decisions are made (RBDM). Key problem solving and decision points become risk-control points. At each of these points, a risk assessment can be conducted to ensure problems are solved effectively and decisions are made correctly.

Depending on the supply chain's RBPS and RBDM efficiency and effectiveness, the chain can be redesigned based on the RBPS and RBDM. Decision process redesign usually follows a 'gap analysis' of mapping the current RBPS and RBDM 'is' processes and developing a vision of the 'should be' SCRM RBPS and RBDM processes. The goal of the process redesign is to move 'is' processes to 'should be' mature RBPS and RBDM processes. The speed by which this is done determines if the change is a transformation or an evolutionary change.

ESTABLISH SUPPLY CHAIN RISK ORGANIZATIONAL STRUCTURE

SCRM requires a coordinated structure to plan and deploy SCRM projects, specifically:

- SCRM corporate steering group.

- SCRM business unit steering group.

- Supply risk innovation teams.

SCRM Corporate Steering Group

This senior level group focuses on the strategic direction of the SCRM initiative at the Enterprise Level. The Chief Supply Management Officer (CSMO) often chairs this group. As well, this group:

- Establishes the vision at the Enterprise Level.

- Develops SCRM guiding principles.

- Provides SCRM resources.

- Establishes internal and supplier rewards and recognition.

- Establishes SCRM metrics/objectives.

- Identifies SCRM process owners in business units.

SCRM Business Unit Steering Group

This operational management group focuses on the tactical direction of the SCRM teams at the Programmatic/Project/Process Level. As well, this group:

- Analyzes internal and customer-supply risk-control systems and supplier processes at the Programmatic/Project/Process Level.

- Identifies specific supply chain innovation opportunities at the Programmatic/Project/Process Level.

- Establishes SCRM plans for achieving objectives and capitalizing on opportunities.

- Provides resources for Programmatic/Project/Process Level initiatives.

- Develops innovative SCRM tactical plans.

- Measures and reports on SCRM tactical progress.

- Provides operational SCRM coaching and mentoring.

- Develops SCRM project leaders and RBPS/RBDM process facilitators.

- Monitors internal and supplier processes for risk-control and capability.

- Intervenes when required and if SCRM objectives are not being met.

Supply Risk Innovation Teams
These teams implement innovation project using established 4P's, RBT, RBPS, and RBDM methodologies and techniques at the Transactional/Product Level. These roles and responsibilities are detailed in the SCRM Deployment – Certification and SCRM Deployment – Supply Selection chapters.

KEY ELEMENTS IN THIS STEP OF THE SCRM JOURNEY
SCRM factors to consider in this step include:

- Design and deploy a risk management process and risk management framework.

- Define who has the authority and accountability to implement internal risk-controls and recommend risk treatment.

- Ensure value creation (upside risk) and waste reduction (downside risk) are understood and pursued by critical supply stakeholders.

- Establish internal consistent, scalable, documented, repeatable risk-control processes for the SCRM initiative.

- Ensure adequate resources are allocated to ensure proper deployment of internal risk-controls.

- Design and deploy suitable and reasonable risk assessment processes.

- Streamline, automate, and lean (if possible) supplier and product work flows.

- Understand how fulfilling end-product manufacturer SCRM requirements assist suppliers and key stakeholders in measurable ways.

- Develop SCRM performance, quality, risk, or other metrics/objectives for suppliers.

- Flow chart core internal and customer-supply SCRM product processes.

- Flow chart internal and customer-supply SCRM RBPS and RBDM processes.

- Streamline supplier and product work flows.

- Identity product choke points and mitigate risks.

- Identify RBPS and RBDM choke points and mitigate risks.

- Assure critical supply chain processes are stable, capable, and improving.

- Use heat maps, risk dashboards, Statistical Process Control (SPC) charts wherever possible to enhance SCRM reporting.

- Reduce risk exposures by installing 4P's controls.

- Ensure information flows quickly to critical supply chain stakeholders.

- Ensure customer-supply risk points are identified with sufficient controls to minimize risk within the end-product manufacturer's risk appetite.

- Ensure adequate resources are allocated for information security.

STEP 4: IDENTIFY SCRM OBJECTIVES, KPI'S, AND KRI'S

The line between disorder and order lies in logistics…
Sun Tzu

More often, supply chain innovation initiatives or projects are performance driven. Demonstrable results are tied to strategic business and SCRM objectives. SCRM innovation efforts are linked throughout the end-product manufacturer, across functional boundaries, and into the supply stream.

As we have discussed, the supply chain consists of a primary process with sub-processes. Critical process elements have innovation goals and SCRM objectives. The chain is only as strong as its weakest link. As much as possible, the supply process chain is standardized, simplified, documented, stabilized, made capable, measured, and improved. Value adders are reinforced. Value detractors, such as waste, are eliminated. These can only be done if there are realistic measures.

SCRM GOALS AND OBJECTIVES

SCRM objectives start at the top of the end-product manufacturer's organization. Corporate SCRM goals are first established and then are deployed to employees and key supply-partners. SCRM goals and objectives are then established throughout the supply chain. Supplier daily, weekly, monthly, quarterly, and yearly activities are measured towards meeting these objectives.

SCRM MEASURES ARE BUSINESS MEASURES

The CSMO usually leads the SCRM and strategic sourcing discussion with executive management. However, operations managers can demonstrate and explain how the supply chain impacts the bottom-line.

Increasing supplier performance is now considered part of the competitive-ness equation. Five years ago, supply risk measures were not on the radar screen. Now, the Board and executive management want to know that objectives, KPI's and KPI's are met.

Key Risk Indicators (KRI's) may involve:

- **Financial measures.** Financial measures are the most critical sup-ply risk measures. KRI's such as income, return on equity, and growth in book value indicate the supplier can fulfill its obligations, invest for growth, and reduce debt through strong assets and strong cash flow. Operations managers must demonstrate how a streamlined supply chain improves business financial measures.

- **Shareholder value.** SCRM's contribution to end-product manufac-turer value is another important indicator to executive management of a company's competitive health. Shareholders expect a return consistent with their expectations and risk. If they do not receive the returns they expect, they will move their assets to another area. Unfortunately, this can sound like a Darwinian message to manage-ment, to continually appreciate stock value, dramatically cut supply chain costs, develop products quickly, and meet ever increasing fi-nal-customer expectations

- **Percentage of total sales.** Total or increasing sales indicate that company is growing. The supply chain is expected to contribute to this in demonstrable ways. Sales outside the U.S. are an important factor for many businesses because domestic markets are mature and exhibit slow growth.

- **Final-customer satisfaction.** Final-customer drives the value chain. If the final-customer product and service experience is highly rated, then the value chain continues to generate value and income.

- **End-product manufacturer satisfaction.** End-product manufac-turer satisfaction delivered through the supply chain is another im-portant measure of a supplier's performance. Studies indicate that end-product manufacturer satisfaction and supplier profitability tend to be positively correlated.

- **Market entry.** New market expansion indicates the company is searching for new growth opportunities. With reduced lifecycles, stagnant markets, and competitive pressures, executive management expects SCRM initiatives to lower costs and facilitate growth in new markets.

- **Market diversification.** Diversification of supply chain risks minimizes market, competitive, and end-product manufacturer risks.

- **New product development.** High-quality, new products, and services are the foundation of a company's continued profitability. Time-to-market is an essential strategic metric to SCRM. Without new products, a company's competitiveness withers.

- **Core competencies.** A company's core supply chain strengths allow it to lever its ability to develop new products, enter new markets, and develop new applications.

- **Reputation.** Perceptions of quality, excellence, profitability, and efficiency are sometimes just as important as their reality. So, end-product manufacturers enhance their brand reputation through consistent advertising, promotions, and public services up and down the supply chain.

KEY ELEMENTS IN THIS STEP OF THE SCRM JOURNEY
SCRM factors to consider in this step include:

- Build out risk-control models at the 1. Enterprise level; 2. Programmatic/Project/Process level; and 3 Transactional/Product level for supplied products and services.

- Use the 4P's model to avoid risks, focusing risky suppliers and commodities.

- Identify and manage supplier concerns proactively, preventively, predictably, and preemptively (4P's).

- Define risk appetite at the Enterprise level and risk tolerance at the Programmatic/Project/Process and Transactional/Product levels.

- Look at the entire supply chain not just parts of the supply chain (s).

- Enforce supplier compliance but focus on higher levels of risk management maturity.

- Establish a set of roles and responsibilities for SCRM that ensures appropriate stakeholders are involved in decision making (RBDM), including identifying who has the required authority to act, who has accountability for an action or result, and who is consulted and/or informed.

- Analyze risks in terms of Black Swan events disrupting the supply chain.

- Analyze risks that may impede reaching SCRM objectives or KRI's.

- Assess each event risk and risks impeding reaching SCRM objectives in terms of likelihood and consequence.

- Conduct supplier audits including compliance, systems, product, process, programmatic, and even forensic audits of high risk suppliers.

- Evaluate supply-partner's state-of-the-art capabilities.

- Use multidisciplinary SCRM teams to improve processes and develop products.

- Use supply certifications to establish a baseline for supply selection and for benchmarking performance.

- Prioritize critical supplied products and services items based upon a common risk assessment methodology such as a heat map.

- Deploy a risk management architecture and processes.

- Establish consistent, well-documented, repeatable processes for determining consequence levels.

- Develop risk assessment processes after the consequence level has been defined, including criticality analysis, threat analysis, and vulnerability analysis.

- Deploy quality and reliability (product risk) program that includes risk-assurance, quality assurance, and quality control process and practices.

- Monitor supplier performance based on compliance, governance, risk, and other types of exposures.

- Evaluate past performance of critical suppliers.

- Develop Information Technology (IT) processes to facilitate comprehensive and real-time reporting of suppliers' performance.

- Ensure that adequate resources are allocated for information security.

STEP 5: DEPLOY SCRM PROJECTS

Standards should not be forced down from above but rather set by the production workers themselves.
Taiichi Ohno

An end-product manufacturer starting the SCRM journey will often focus on projectizing work. The focus usually starts with commodity suppliers, where risks are manageable. The project focuses on low hanging fruit, where success is visible and quantifiable. The project approach allows for lessons learned to be replicated and scaled to larger suppliers.

The basic steps in a SCRM project for an end-product manufacturer and supplier include:

- **Identify and prioritize supply chain innovation opportunities.** Opportunities are identified and prioritized through Pareto analysis (80-20 rule) of the cost of quality, product lifecycle costs, return on investment, risk analysis, and force field analysis. SCRM projects with the highest return, quickest return, or highest risk are chosen and pursued first.

- **Qualify and certify key suppliers.** Develop processes for qualifying suppliers and managing the supply portfolio; establish clear standards and specifications; evaluate in-depth background information on supplier financial stability; reveal compliance standards; and supplier's business ethics.

- **Review overall sourcing risks.** Identify designs, tests, quality controls, ingredients, components, and equipment required to produce products or deliver services. Identify any sourcing risks and implement controls, policies, and processes to eliminate avoidable risks and mitigate unavoidable risks.

- **Understand supplier's risk-controls and contingency plans.** Confirm key suppliers have solid risk remediation and business continuity programs, including redundant supply, transportation, equipment, and infrastructure.

- **Maintain and review supplier product and service-level agreements (SLAs).** Establish realistic service level agreements (SLA's) with financial incentives for on-time deliveries of quality products and appropriate penalties for sub-performance.

- **Collaborate with critical suppliers.** Leverage network collaboration across suppliers to optimize supply relationships; optimize supplier assets; review spend; reduce costs; negotiate improved service levels; and achieve improved operational (time, quality, and cost) efficiencies.

- **Review supplier equipment, resource, and packaging processes.** Identify key manufacturing equipment to ensure appropriate level of process control; review testing and laboratory needs; assure appropriate environmental infrastructure with backup; ensure supplier systems with critical dependencies that have a high impact or potential for failure; develop supplier wide plan to manage equipment and systems, suppliers, spare parts, and equipment maintenance and repair capabilities. Establish SLA's to confirm adequate response times for mitigating potential disruptions.

- **Flow chart (map) supply chain processes.** Before a supply innovation project is initiated, supply chain processes are flowcharted for resource, problem solving, and decision flows. A supply flowchart shows the process chain as a series of steps or links. Each step has a customer and supplier. Critical problem and decision points are risk points that are risk assessed. The flowchart can then be used to understand the supply process and to identify redundancies, waste, or other non-value-added activities. The objective is to pursue lean, quality, or waste initiatives. Specific techniques used to flowchart processes include: block diagrams, input/output analysis, benchmarking, and process redesign.

- **Assess fulfillment of final and internal customer needs.** Each process step has a customer whose needs are satisfied. Customer satisfaction determines if the process is doing what it is supposed to be doing.

- **Develop and establish supply chain process measures.** The inputs, process, and outputs of the process are measured. Measures can be end-product manufacturer, process or product specific. For example, process measures may look at speed, quality, cost, delivery, efficiency, or effectiveness.

- **Understand sources of supply process variation.** The measurement and control of process variation is key to all customer-supply innovation. Once data are collected and reviewed, then special causes of variation can be eliminated. Once the process is in control, then common or fundamental causes of variation can be identified and eliminated.

- **Control process variation.** Supply chain processes can then be controlled, standardized, and proceduralized. When a supply process has been stabilized, then it can be improved.

- **Create redundancy plans.** Develop design, equipment, production, people, and technology redundancies for critical manufacturing that may not have back-ups. Identify alternative equipment to be validated for production to provide back-up and reduce recovery time should a piece of equipment malfunction.

- **Confirm adequate stock of supplied key parts.** For key products and supplies, ensure safety stock levels are sufficient to maintain supply continuity in the event of a supply disruption.

- **Establish safeguards against product counterfeiting and diversion.** Policies, procedures and security safeguards are designed and deployed to prevent the end-product manufacturer's, supplier's designs, or brand owner's products from being counterfeited or diverted from the traditional supply chain. For example, authentication technologies and packaging technology like security codes can be designed to minimize the impact of counterfeit products such as drugs. Key product shipments should be fully traceable from the source even if the product has been repackaged and relabeled by using Radio Frequency Identification (RFID) tags.

- **Integrate new regulations into the SCRM risk and compliance program.** Develop policies and procedures for supplier to end-product manufacturer traceability; understand and account for regulatory-driven requirements into operational and budgetary plans; plan for personnel, capital, and other resources needed to meet evolving regulatory requirements; integrate requirements into sourcing; develop SLA's and contract management; maintain continuous end-to-end process monitoring of supply chain risks; initiate immediate changes if risks are discovered; partner with strategic alternate suppliers who can provide back-up products or services.

KEY ELEMENTS IN THIS STEP OF THE SCRM JOURNEY
SCRM factors to consider in this step include:

- Scope and list possible SCRM suppliers, authorized distributors, and certified resellers.

- List possible SCRM projects with ROI attached to each.

- Selected simple and high ROI projects to pursue at the 1. Enterprise level; 2. Programmatic/Project/Process level; and 3 Transactional/Product level for supplied products and services.

- Deploy consistent, well-documented, repeatable processes for risk engineering, reliability engineering, system engineering, process engineering, security practices, and product acquisition.

- Establish internal checks, risk-assurances, and balances to ensure compliance with quality, delivery, cost, and risk requirements.

- Reduce the supply base, sometimes to single-source partners based on context and risk appetite.

STEP 6: EVALUATE SCRM PROJECT RESULTS

If you think of standardization as the best that you know today, but which is to be improved tomorrow; you get somewhere.
Henry Ford, founder Ford Motor Company

From a business perspective, measurement determines if SCRM investments help assure meeting SCRM objectives. Objectives may involve lowered costs, reduced deficiencies, reduced cycle times, improved productivity, improved customer satisfaction, or other specific factors.

SCRM Programmatic/Project/Process and Transactional/Product KRI's and objectives may include:

- **Process measurements.** Process measures include final-customer satisfaction, internal customer satisfaction, supplier process control, and supplier process capability.

- **Project measurements.** Project measurements involve meeting internal SCRM budgets, completing SCRM projects on time, and complying with end-product manufacturer contractual requirements.

- **Product measurements.** Product measurements include complying with end-product manufacturer's dimensional, physical, or chemical specifications.

KEY ELEMENTS IN THIS STEP OF THE SCRM JOURNEY
SCRM factors to consider in this step include:

- Develop end-to-end understanding of processes, resources, monies, schedules, costs, quality, and other supply risk factors.

- Measure gaps in KRI, KPI, and objective performance.

- Remediate and/or correct performance gaps.

- Monitor supplier performance based on compliance, government, risk, and other types of exposures.

- Conduct supplier audits including compliance, systems, product, process, programmatic, risk, and even forensic audits if it entails high risk suppliers.

- Identify and manage supplier concerns proactively, preventively, predictably, and preemptively (4P's).

- Deploy an appropriate and tailored cyber security controls.

- Develop Information Technology (IT) processes to facilitate comprehensive and real-time reporting hose supplier this performance.

- Design and deploy a tested and repeatable business continuity plan to ensure operability and reliability of the supply chain following a 'Black Swan' event.

STEP 7: REVIEW AND START AGAIN

Leaders get out in front and stay there by raising the standards by which they judge themselves – and by which they are willing to be judged.
Frederick W. Smith, founder of FedX

One of W. Edwards Deming's (the quality guru) 14 points of quality management was 'constancy of purpose' which meant that process innovation was not instant pudding but required continuous attention and application. SCRM requires long-term commitment and effort from executive management and day-to-day commitment from each supply and organizational stakeholder.[121]

CONTEXT: Critical Success Tips for Supply Chain Resilience

Problem: First-tier suppliers' risk are unknown.
Solution: Understand first-tier suppliers' risks and how they are managed. As mitigation, look at supply chain diversification.

Problem: Poor supplier communications and collaboration.
Solution: Be open with suppliers from the outset. As mitigation, communicate SCRM requirements and expectations.

Problem: Transactional approach with key suppliers including punishment for noncompliance.
Solution: Take the long-term view. SCRM is a new way of management. As mitigation, provide incentives to suppliers to work with the end-product manufacturer.

Problem: SCRM by gut.
Solution: Do the analysis. SCRM is not reactive management. It is RBPS and RBDM. This takes time. As mitigation, develop a SCRM plan.

Problem: Focus on short-term delivery and quality.
Solution: Evolve approach through lessons learned. We live in VUCA time. Risks will increase and mistakes will be made. As mitigation, learn from mistakes.[122]

Continuous supply chain innovation implies the process starts again when a project has been completed and a process has been stabilized. The innovation cycle may start with another supplier, expand to become a SCRM initiative or become a targeted supply innovation project.

An important element of this step is to review the status of the SCRM journey, then start over. Many projects tend to be a one-off. The project finishes and the SCRM initiative languishes. McKinsey captured this risk:

> "Many organizations have multiple independent initiatives underway to improve performance, usually housed within separate organizational groups (e.g. front and back office). This can make it

easier to deliver incremental gains within individual units, but the overall impact is most often underwhelming and hard to sustain. Tangible benefits to final-customers - in the form of faster turnaround or better service - can get lost due to hand-offs between units. These become black holes in the process, often involving multiple back-and-forth steps and long lag times. As a result, it is common to see individual functions reporting that they have achieved notable operational improvements, but customer satisfaction and overall costs remain unchanged."[123]

KEY ELEMENTS IN THIS STEP OF THE SCRM JOURNEY

SCRM factors to consider in this step include:

- Develop business continuity and contingency plans for 'Black Swan' events and not meeting SCRM objectives.

- Establish SCRM reporting to the right stakeholders with the required information at the right time.

- Design supplier monitoring and risk-assurance based on enterprise risk appetite and supplier capabilities.

- Implement a tested and repeatable contingency plan that integrates SCRM considerations to ensure the integrity and reliability of the supply chain including during adverse events (e.g., natural disasters such as hurricanes or economic disruptions such as labor strikes).

- Deploy a robust incident management program to successfully identify, respond to, and mitigate security incidents. This program can identify causes of security incidents, including those originating from the supply chain.[124]

RESILIENT SUPPLY CHAINS

During the last war, eighty percent of our problems were of a logistical nature.
Field Marshall Montgomery

Supply disruption is emerging as a top risk-control concern for end-product manufacturers. In a recent survey, supply disruption doubled in importance relative to other enterprise concerns. Specifically, 48% of companies are concerned or extremely concerned. Also, 75% of the surveyed companies had at least one recent supply disruption.[125]

The reality is that most end-product manufacturers are not resilient in terms of a major supply chain disruption. Why? Many companies do not analyze the root-cause of a supply chain disruption according to recent surveys.[126]

'WHAT IF' QUESTIONS

Supply chain disruptions will occur. It is not a matter of if but when. SCRM can often be boiled down to asking the right RBPS and RBDM questions at the right time. One critical risk question to ask is 'what if'? If the disruption occurs with a specific supplier, what are the possible consequences.

The question can help an end-product manufacturer visualize potential risks in the supply chain and eventually quantify and prioritize them. Another way to say this is to ask what can go wrong in the supply chain, whether it involves suppliers, processes, and products.

When this is determined, then critical questions can be asked such as what is the consequence to the business if this disruption occurs? What can the brand owner or the end-product manufacturer do to prepare for the possible disruption? This can lead to thinking about supply chain resilience; scenario thinking; and developing business impact analyses and contingency plans.

SUPPLY CHAIN RESILIENCE

SCRM emphasizes business continuity and resilience. Business continuity management is the ability to ensure that severe supply chain disruptions will not break the supply chain. Business continuity management usually starts with business impact analysis. In business impact analysis, threats, and vulnerabilities are first identified to determine how they can impact the business or key processes if they are disrupted.

Once impacts are identified, then supply chain and business resilience can be determined. What does supply chain resilience mean? It is an interesting concept that incorporates risk solutions, such as business continuity, business impact analysis, risk management, process consistency, and disaster recovery.

Resilience in terms of the entire end-product manufacturer implies the event or threat was detected sufficiently early so the end-product manufacturer or the supply chain is not totally disrupted.   Resilience may involve securing:

- **Multiple suppliers.** Obtain additional production or distribution facilities.

- **Different logistics.** Determine different routing or choosing different forms of transportation for goods.

- **Sharing tactics.** Develop sharing strategies with a supplier to mitigate risks.

- **Additional inventory.** Build inventory redundancies as a buffer against uncertainty.

BUSINESS IMPACT ANALYSIS

Business impact analysis (BIA) identifies key business systems, processes, information, authorities, people, financials, and resources that are needed to maintain operations with the end-product manufacturer. A supply disruption can result in business costs that can include: final-customer dissatisfaction, loss of revenue, reputation loss, interrupted revenues, higher internal costs, or possible contractual penalty payments.

The BIA analysis can be a short document or it can be very detailed reviewing all SCRM at risk systems and vulnerable business processes. For each key critical business system, the BIA will review how a disruption could impact the end-product manufacturer.

A SCRM plan is developed preferably before a supply chain incident or disruption. By identifying the key business areas possibly impacted, the end-product manufacturer can prioritize and focus its 4P's plan (proactive, preventive, preventive, and preemptive) on critical areas.

The breadth and depth of the impact on the business depends on the type, financials, length, and severity of the supply chain disruption. The disruption can impact the end-product manufacturer or brand owner at the:

1. **Enterprise level.** End-product manufacturer would have to stop operations due to the disruption.

2. **Programmatic/Project/Process level.** End-product manufacturer would issue a corrective action request or issue a request for root-cause analysis.

3. **Transactional/Product level.** End-product manufacturer would contain the defective or nonconforming products, issue a request for additional products from the supplier, and initiate a corrective action.

SCRM CONTINGENCY PLAN

Prevention and preparation can assure an acceptable level of business continuity if the supply chain is disrupted. Depending on the risk appetite of the end-product manufacturer, the SCRM contingency, incident response, the business continuity plan can be a detailed document addressing many 'what if' questions and possible disruptive scenarios.

It is recommended the end-product manufacturer focuses on the most critical supply vulnerabilities and develop contingency plans for disruptions with critical first-tier suppliers. Otherwise, the end-product manufacturer will be boiling the ocean trying to develop plans for all disruptive contingencies.

Companies with mature lean programs still develop contingency plans for 'what if' scenarios. Airbus, one of Europe's biggest end-product manufacturers, has lean operations with few inventories. The challenge is the external risk context for Airbus changed with Brexit, the U.K's move out of the European Union. Airbus builds its planes in 4 countries across Europe including the U.K. This has created an external context risk: "Airbus ... has warned it may have to stockpile parts to operate smoothly once the U.K leaves the E.U."[127] The risk is that Airbus has a just-in-time supply chain

---

CONTEXT: Tips for Establishing a SCRM Initiative

- Take a 360^0 view in defining enterprise level supply risk.
- Continually update SCRM plans to mirror the shifts in the supply base and marketplace.
- Develop ongoing SCRM training programs.
- Establish cross-functional teams to support the SCRM program.
- Expect the best, prepare for the worst.
- Understand that SCRM does not mean the elimination of supply risk.[128]

---

and that a three-hour delay of materials from the U.K. to Europe would not be an acceptable risk.

BUSINESS CONTINUITY PLAN

The SCRM business continuity and response plan determine what is needed to manage efficiently, economically, and effectively. The immediate challenge from a supply chain disruption perspective is getting operations back on line. In general, the longer it takes to restore normal operations, the greater the impact and cost to the business.

Disaster Recovery is the ability of an end-product manufacturer to be resilient if there is a disruptive event. Every company is vulnerable at critical risk points of the supply chain. Some points or nodes are more critical than others. If these nodes are disrupted, then the supply chain will become unstable and possibly stop. The goal is to develop and design a resilient supply chain so that if an event occurs, business operations can be restarted quickly.

Some end-product manufacturers have established business continuity and crisis management teams to ensure that most forms of disruption in the supply chain are manageable. In some ways, this can also be impractical. Why? There are simply too many unknowns. Companies can spend a lot of time looking at risk-controls, underlying risk assumptions, and critical flows of their supply chains without receiving value.

On the other hand, an end-product manufacturer may not realize the extent or cause of a possible disruption such as a cyber breach because it is perceived as a minor nuisance. End-product manufacturers may not be aware of the threats and risks they face especially beyond first-tier suppliers. In other words, the end-product manufacturer is not aware of the direct and indirect consequences of a disruption that has not occurred. End-product manufacturers often do not have actionable information to develop commensurate controls of the possible risks. Finally, global supply chains may be exposed to Black Swan risks that have unknown or unknowable cascading impacts.

RISK-ASSURANCE AND RESILIENCE

Resilience can mean many things. Resilience may imply business continuity. Resilience may include the agility to respond to possible disruption with backup processes and products placed in inventory. Resilience may mean minimal impact on customer satisfaction and to the organization's financials.

Can supply chain risks be managed with complete certainty? No. There is no such thing as absolute certainty and 100% risk-assurance. The best an end-product manufacturer can attain is probably some level of resilience based upon its risk appetite.

FINAL THOUGHT - SOCIALIZING SCRM

A hard lesson learned dealing with SCRM and ERM is the technology of SCRM is straightforward. The behavioral changes in the socialization that are required to architect, design, deploy, and assure the SCRM program require much time and effort.

SCRM requires a coordinated effort across the end-product manufacturer's organization and across its supply chains. Building trusting relationships with suppliers, stakeholders, end-product manufacturers and interested parties requires time. This is what we call socializing SCRM. Socializing risk management is critical for its adoption and ultimate success.

SUMMARY

Supply chain management traditionally has been reactive. If an event, threat, or obstacle appeared, the end-product manufacturer would have all-hands eliminate the risk or threat. This was reactive supply management.

SCRM adoption requires a culture shift. Operations managers are encouraged and rewarded to learn, adapt, and adopt new technologies and behaviors. The best SCRM processes are built to encourage agility, quickness, and compatibility. Products are developed quickly. Virtual supply-partners are selected. A culture of rapid innovation and SCRM development is emphasized daily.

Resilient companies have a culture that thrives on VUCA change, specifically technology, customer, system, competitor, and marketplace disruption. The faster the rate of change, the more these end-product manufacturers thrive competitively. How is this done? Cycle times are reduced. New value-added technologies are rapidly deployed. These end-product manufacturers use change to maintain their competitive edge and to enhance profitability. It is not a matter of being left behind but anticipating final-customer expectations and leapfrogging the competition. SCRM companies quickly develop, test, and tweak new technologies, core processes, and organizational structures to be flexible and competitive.

NEXT CHAPTER
In the next chapter, we discuss certification, a critical element of SCRM deployment.

CHAPTER 10:
SCRM DEPLOYMENT –
CERTIFICATION

WHAT IS THE KEY IDEA IN THIS CHAPTER?

End-product manufacturers now consider SCRM critical to success and competitiveness. More than 76% of executives consider SCRM an important or very important strategy.[129]

We have discussed a risk management framework, process approach, and project approach to SCRM design and deployment. In this chapter, we discuss supply certification as another approach to SCRM. Supply certification is based on improving suppliers' capabilities and maturity.

Is there one right method? Any approach will work depending on organizational objectives and context. The key idea in this chapter is that supply certification can be used early or even as the first step in SCRM.

CUSTOMER-SUPPLY CERTIFICATION

All business proceeds on beliefs or judgments of probabilities, and not on certainties.
Charles Eliot, President Harvard University

In the last chapter, we outlined the general steps in the SCRM project journey. They do not have to be followed sequentially. The critical point is to determine a starting point for SCRM and start the process.

REVISITING VUCA

A key element of SCRM planning is to understand supplier VUCA and the risk assumptions for selecting a supplier. In many cases, the basic decision for selecting a supplier was made 5 or even 10 years ago and the basic assumptions that lead to supply selection now need to be revisited due to increasing VUCA risks, such as:

- **Volatility.** Volatility is increasing in terms of spikes in fuel prices, increases in raw materials costs, or even changes in customer demand.

- **Uncertainty.** "Uncertainty has become a way of life for business leaders since the last financial crisis. The unprecedented turbulence is seen in the deep political divisions in the U.S. and abroad, uncertainty over the strengthened U.S. dollar, the faster pace of technological change and looming global problems such as mass migration, terrorism, and climate change (RBPS)."[130]

- **Complexity.** Everything seems becoming more complex such as supply chains, IoT products, software, cyber security, and even sourcing business models. Complexity inherently increases the probability of a disruption, product nonconformances (defects), and other supply chain risks.

- **Ambiguity.** Suppliers and supply chain issues that seemed straight forward, now seem less certain. Reliable suppliers start shipping nonconforming products. Suppliers change components without notifying the end-product manufacturer (RBDM).

WHERE TO START?

An end-product manufacturer will have suppliers in different product and service categories all of which are at different levels of maturity and capability. We recommend that the end-product manufacturer starts the journey with one set of simple product suppliers such as commodity suppliers. These suppliers will probably have similar risk maturity and capabilities. A supply certification program involving risk with these suppliers will provide the end-product manufacturer with experience and lessons learned that can be used with suppliers of more technical products and sophisticated services.

Increasing a supply-partner's maturity and capability should offer mutual wins. The supplier gets a heads up on what is coming down the product

CONTEXT: Benefits and Disadvantages of Certification Standards

The benefits of certification standards include:

- Compatibility.
- Interchangeability.
- Intercommunications.
- Trade facilitation.
- Technology transfer.

The disadvantages of certification standards include:

- Slow to develop.
- Require consensus among many stakeholders.
- Are sometimes obsolete by the time they are published.
- Are often descriptive, instead of prescriptive.
- Do not represent the ideas of all stakeholders.
- Are difficult to decipher.
- Do not relate to daily business concerns.
- Are difficult to locate

development pipeline and is in a favorable sourcing position when contracts are bid. Or, the supplier's early involvement can lead to preproduction contracts that can lead to cycle time, cost, and schedule reductions. The final-customer wins because the end-product manufacturer has the benefit of the supplier's core capabilities.

SCRM process often starts with product/service requirements planning as the following describes:

> "Buyers have become increasingly dependent on formal methods and procedures for ensuring that the products, services, and systems they purchase, whether domestic or foreign, consistently meet their needs. However, not all product and service characteristics can be evaluated simply by picking up and examining the item in the marketplace. Such characteristics need to be determined and assessed, providing confidence to the buyer (or other interested

party) that the product conforms to requirements and that conformance is consistent from product to product.[131]

CUSTOMER SUPPLY CERTIFICATION

End-product manufacturers establish formal supply certification programs to induce suppliers to improve. Suppliers are classified into categories based on SCRM performance, maturity, and capability. Partnering levels include candidate, approved, preferred, and supply-partner. ISO 9001:2015 certification with its RBT requirements is a good first step to achieve 'candidate' status. Once selected as a candidate supplier, the supplier is expected to improve its product quality, service, delivery, cost, and risk processes.

Supply-partners, the highest certification category, usually have in-house quality, cost, logistical, lean, JIT, or other processes that are equal or more stringent than the end-product manufacturers. Supply-partners are audited periodically to ensure a supplied product is defect-free. Preferred supplier parts are also sent directly into production as part of the lean, just-in-time production system.

SUPPLY CHAIN METRICS/OBJECTIVES

Success, as I see it, is a result, not a goal.
Gustave Flaubert, Writer

Metrics/objectives are the basis of SCRM certification. As a supplier meets, achieves, and then surpasses specific metrics and objectives, the supplier moves up the customer-supply certification ladders. The metrics and objectives verify and validate supplier performance. The metrics and objectives can be used to select, improve, correct, and reward suppliers. Often, candidate and approved suppliers must meet compliance criteria or thresholds such as being certified to ISO 9001-2015 or they can be evaluated on numerical criteria. More often, ratings are numerical based on a 100 or even a 1000 point score.

INTERNAL SCRM METRICS/OBJECTIVES

In the last several years, SCRM objectives were only incorporated in the compensation and performance plans of senior supply management. Since SCRM is an all-inclusive function, we are seeing materials, quality, planning, engineering, and manufacturing managers have SCRM performance measures. One driver of this trend is to bring internal organizational silos and key suppliers closer together.

Measurement is essential to SCRM development. Without complete and accurate data, decisions (RBDM) cannot be made of a supplier's performance. Supply metrics/objectives are designed and tailored to the maturity of the supply chain and to the capability of the specific supplier.

Many SCRM KRI's are process and product based, specifically dealing with manufacturing process control. What does this mean? Supplier teams and process personnel attempt to proceduralize and stabilize operations that are capable of satisfying final-customer requirements. If there is an unusual occurrence, the supply process is stopped and the source of the problem is discovered and eliminated. Control depends on process owners using their best judgment to make the right decisions to solve the problem (RBPS).

EXAMPLES OF SUPPLY PERFORMANCE METRICS/OBJECTIVES
Common measures used to track and monitor suppliers include:

Supply Quality Metrics/Objectives
Examples of supply quality measures include:

- Process capabilities.

- Product quality performance.

- Production line performance.

- Reliability performance.

- Final-customer returns.

- Corrective action responsiveness.

- Parts per million quality levels (Six Sigma).

- Rejects of incoming material.

CONTEXT: Supply Risk Measurement Benefits

- Convey final-customer requirements and inducements throughout the supply chain.
- Establish agreed-upon means to design SCRM.
- Establish supply baselines for quality, cost, and delivery variances leading to risk.
- Help to develop supplier KRI'S, performance benchmarks, and opportunities for improvement.
- Create risk mitigation or upside risk opportunities (RBPS/RBDM).
- Assist both end-product manufacturer and supplier to improve risk performance through a variance gap analyses of delivery targets, costs, quality, and technology.
- Identify significant variances (risks), then risk-controls are developed.
- Identify customer-supply quality, cost, risk, and schedule trends.
- Identify the best suppliers to work with and to move up the risk maturity and capability curve.
- Identify where to commit limited resources.
- Is a means to evaluate the overall supply chain risk efficiency, economics, and effectiveness.

- Rework in dollars or hours.

- Production stoppages.

- Final-customer complaints.

- Recurring nonconformances (product defects).

- Amount of material accepted on-waivers.

- Number of late shipments.

- Tighter specifications.

- Technical competency.

Supply Service Metrics/Objectives

Examples of supply service measures include:

- Proven service leadership.

- Willingness to partner with end-product manufacturer.

- Pricing and cost reduction commitments within end-product manufacturer expectations.

- Favorable payment terms.

- Availability of products.

- Flexibility in schedules.

- JIT processes in manufacturing and material control.

- Local inventory points.

- Ease of doing business and communications.

Supply Support Metrics/Objectives

Examples of supply support measures include:

- Help desk capabilities.

- Design methodology.

- Product engineering/development support.

- Early supply involvement.

- Concurrent engineering commitment.

- Product time-to-market development schedules.

- Supply quality assurance support.

- Supply open and honest communications.

- Service development responsiveness.

Supply Financial Metrics/Objectives

- Economic value-added.

- Total cost reductions.

- Scrap costs generated in use.

- Warranty costs due to product failure.

- Cost of doing business in the development process.

- Sourcing life cycle cost reduction.

- Maintenance costs reduction.

- Equipment warranty cost reduction.

- Disposal costs.

- Sourcing transaction cost reduction.

- Inventory costs reduction.

- Equipment uptime and share cost reduction.

BALANCED SCORECARD

SCRM scorecards are often used to evaluate supplier performance. The idea behind balanced scorecards is that end-product manufacturer and supplier interactions are graded based on a mutually established measurement method and based on product criticality. For example, high-value goods and services would be evaluated by different value attributes and weighted differently than commodity items.

Then throughout the contract or product lifecycle, transaction events, processes, products, or shipments are evaluated based on process criteria and product value attributes. Does this mean the end-product manufacturer develops a new set of standards or value attributes for each supplier? No. However, the end-product manufacturer and supplier should negotiate value attributes based on the product risks.[132]

How effective is this method? Traditionally, the end-product manufacturer measured supplier performance based on a one-way scorecard. The end-product manufacturer tested these with a critical few or even all suppliers. This sometimes worked great or it was simply seen as window dressing for one-way end-product manufacturer demands upon the supplier. It is now hoped that the new balanced scorecard, like 360-degree human resource performance evaluations, would be fair and balanced.

SCRM BENCHMARKS

If you don't measure it, you can't manage it and it won't happen.
Anonymous

More end-product manufacturers are developing aggressive supply cost, delivery, risk and quality metrics/objectives. For example, Chrysler is expecting 15% cost reduction from suppliers. GE expects Six Sigma (parts per million) quality levels from service and product suppliers. And, the list goes on

INCREMENTAL OR AGGRESSIVE CERTIFICATION

There is a big debate whether SCRM is incremental or aggressive. What we do know is that disruption is accelerating. Continuous process risk innovation is based on a series of singles winning the SCRM ball game. Innovations are regular and focused on things that matter to the end-product manufacturer. Some say, that incremental innovation is necessary but not sufficient in a highly competitive marketplace that rewards first-movers, first-developers, and first-to-critical-mass.

The benefit with incremental innovation is that there are fewer surprises for the end-product manufacturer and the supplier. The counter argument is that incremental innovation can breed relaxation and mediocrity over the long-term. Stretch benchmarks can be used to encourage flexibility, innovation, and performance innovation. The end-product manufacturer can then rate multiple suppliers based on aggressive criteria to reduce the supply base.

RISK BASED PERFORMANCE

Jack Welch, past CEO of GE, was the nameplate leader for many years. What made him so effective? He always focused on performance. One person who worked for Welch said every limo or elevator ride would be an opportunity for a performance review. "A general manager's quarterly and year-to-date profit and loss variance analysis takes minutes. And the variance is measured not against the budget but against what the manager had promised."[133]

We are seeing more aggressive use of accountability. In an unusual move, "one company's chief executive agreed to forfeit a year's salary if the chemicals company did not meet its earnings target" according to a *The Wall Street Journal* article.[134] Would most managers want to put his/her salary at risk?

What it means to put a manager's salary at risk is not lost on shareholders, competition, and suppliers. Usually, entrepreneurs put their salary and future at risk. Now, we are hearing more about 'shared-risk, shared-reward' compensation in more work and SCRM environments. More operations managers and employees are putting real pay on the line and are betting stock options and bonuses on their success - in other words, more operations managers are asked to 'walk their talk.' More than 51% of organizations, tie non-management and non-sales personnel compensation to individual or group performance.[135]

RISK BASE CONTRACTING

Healthcare is increasing its use of SCRM and risk based contracting. What does this mean? Risk based contracts focus on guaranteed performance clauses of healthcare suppliers, including pharmaceutical, medical device, and equipment manufacturers. Performance is measured based on a drug or medical device achieving specific outcomes from clinical trials or a product achieving standards. If the product or service does not or cannot meet the standards, then the end-product manufacturer or provider provides them at a lower price, provides a rebate, or if it deals with a patient implant, covers the cost of a replacement. [136]

CONTEXT: Daimler Chrysler Supplier Metrics/Objectives

Daimler Chrysler management recently wanted immediate action to stem the flood of red ink. Since so much of the manufacturing dollar resided in the supply base, Daimler Chrysler followed a 2-step process. In the first year, suppliers were directed to reduce by 5% the prices for materials and services. In the following years, Daimler Chrysler wanted to reduce costs another 10%.

How did they do this? The additional 10% in cost savings was expected to come from customer-supply partnering arrangements. Customer SCRM teams, representing 75% of Daimler Chrysler's material purchases, would work together to identify cost innovation areas.[137]

This is part of a larger trend. Entitlement is dead. Accountability is alive and well. Everyone in the supply chain is responsible for their activity, process, function, team, plant, and company performance. Performance management is a direct outcome of self-management and self-managed SCRM teams.

Executive management is looking to supply chain process owners to operationalize the critical elements of the strategic vision, mission, plans, objectives, etc. Each operations manager is now expected to know how his or her contributions fit into the organization's overall goals and how they meet end-product manufacturer requirements.

BIG HAIRY AUDACIOUS SCRM GOALS (BHAG)

BHAG - what a strange acronym? BHAG (pronounced bee-hag) stands for Big Hairy Audacious Goals. Collins and Porras in **Built to Last** call super stretch performance goals BHAGs. Examples of BHAGs include IBM's decision to focus on cloud computing and Boeing's decision to build the first jumbo jet, the 777. Both were outstandingly successful and set both firms apart in their respective sectors (RBDM).

BHAGs also identify superior suppliers. What characterizes a BHAG? It is a stretch goal, an opportunity, bigger than life that inspires people to actualize a vision. A BHAG is doable and definitely stretch the current paradigm of doability.[138] Today's operations managers are expected to destroy the current SCRM paradigms, burst today's envelope, and stretch way beyond the norm.

SUPPLY RISK CERTIFICATION

The supply chain stuff is really tricky.
Elon Musk, CEO of Tesla and SpaceX

The rationale for supply risk certification is straightforward. An end-product manufacturer cannot be all things to all people or to all final-customers. An end-product manufacturer with its brand on the final product or service now assumes all the risks including reputation loss if something goes bad with its product or service. One way to start SCRM is through supply certification based on ISO 9001:2015 RBT requirements.

CERTIFICATIONS

ISO 9001 standard was originally a global customer-supply and Quality Management System certification standard. ISO quality documentation and quality management system assessments were used to certify candidate suppliers. As suppliers moved up the maturity curve, end-product manufacturers would conduct process and risk audits that related directly to end-product manufacturer requirements and customer-supply integration.

Global certifications or marks such as ISO 9001:2015 certification and Japan Industrial Standards (JIS) label are still found in customer-supply contracts. Are these certifications or marks credible? Yes. However, they often offer the end-product manufacturer a low-level of risk-assurance so they are often used to select candidate suppliers.

More stories are arising that end-product manufacturers require higher levels of assurance. For example, Kobe Steel and other companies were certified to ISO 9001. However, Kobe Steel was found to have falsified data. It is believed that the ISO Certification Bodies just did not have the time and resources to conduct an in-depth quality management systems assessment. In these cases, end-product manufacturer are developing additional customer-supply requirements.

CONTEXT: Supplier Certification Risk

Japan once had a world-class reputation for quality and lean management, Toyota Production System, Statistical Process Control, and many other world-class production techniques. However within a year, Kobe Steel falsified quality data; Mitsubishi Motors faked vehicle fuel efficiency data; Toyo Tires & Rubber faked data on earthquake materials; Nissan Motors used unqualified staff to inspect vehicles; Asahi Kasei faked building data; and the list goes on.

The hard lessons learned is that reputation loss impacts a company's financials but also a 'Made in Japan' global reputation.[139]

WORLD-CLASS SCRM

As SCRM increases in importance, more end-product manufacturers may want to work with a 'critical few' or 'world-class' suppliers. What makes a 'world-class supplier'? This supplier has core competencies that the end-product manufacturer does not have. In other words, a 'world-class' supplier is one with known and demonstrable processes that complement the end-product manufacturer.

It is especially critical that each supplier in the value chain of an end-product manufacturer is understood. By itself this previous statement seems self-evident to many companies. If one purchases products from a key supplier, then it is assumed the supplier is known and has been vetted by the end-product manufacturer with few risks.

The critical question is: Does the end-product manufacturer know where or from whom do critical suppliers purchase their products and services. So, if possible the ultimate manufacturer should strive to understand the breadth and the depth of possible supply chain risks. In this way, the end-product manufacturer can identify critical decision points, critical control points, bottlenecks, and weak links in the value chain process and take appropriate actions to mitigate risks (RBDM).

SHIFTING PARADIGM

Another challenge is the supply selection paradigm may have shifted. The manufacturer has been focusing on selecting and outsourcing to single-source, 'world-class' suppliers. But, what does 'world-class' really mean? These suppliers hopefully have mature and capable processes to deliver world-class products on time and to the right location. But, do 'world-class' suppliers have mature SCRM processes? Unfortunately, usually not.

This means that end-product manufacturers and brand suppliers must start a SCRM development initiative with their suppliers, which may be some form of supply risk development, certification, or even partnering. Partnering to develop SCRM capabilities allows an end-product manufacturer to bring key suppliers into its product development process to share its SCRM skills and knowledge.

Key suppliers are offered inducements to embrace supply risk certification. Suppliers can be induced through customer-supply partnering, joint ventures, or strategic alliances so they can penetrate new markets, provide a local presence, enhance product quality, lower risk exposures, create a market for new products, establish a local distribution network, resell products, integrate suppliers, and share process innovations.

TYPES OF CERTIFICATION?

If you don't measure it, you can't manage it and it won't happen.
Anonymous

Supply risk certification is a supply development process. There are several ways to become certified:

- **Self-certification.** The supplier certifies it meets or can meet end-product manufacturer's risk and other requirements. This is done with non-critical products and services.

- **Second-party certification.** The end-product manufacturer wants a higher level of risk-assurance than self-certification. The end-product manufacturer, the second-party, will conduct an on-site system/process/product audit of the supplier's processes and documentation, which traditionally demonstrated compliance to quality, delivery, risk, cost, and technology requirements, but now include risk requirements. The idea is pretty simple. If the supplier's risk-control processes are in place according to end-product manufacturer specifications, then there is a high level of risk-assurance that the output from the processes, the products, conform to end-product manufacturer requirements.

- **Third-party certification.** In the ISO 9001:2015 world, third-party certification is also called registration. Registrars will audit suppliers to ensure they conform to ISO 9001:2015 or other management system requirements. In this way, end-product manufacturers get a level of risk-assurance that the supplier is capable of meeting requirements by checking a registry of certifications.

TYPES OF SUPPLY CERTIFICATION

There are several basic types of certifications:

- **Product.** Supplied products are tested to verify critical dimensions, performance capabilities, or chemical/physical characteristics. Product testing is often regulatory, ensuring compliance to a standard or regulation. Sometimes, 'first item' testing is 100% testing or inspection of the first product produced from a production process.

- **Management system.** Systems testing usually deals with policy, procedural, and work instruction conformance. ISO 9001:2015 is one of the best-known management system evaluations. Management system certification is usually binary. Documentation addresses and complies with the standard or it does not.

- **Process.** Process evaluations deal with flowcharting a process, following it from beginning to end, checking for risks, and then determining the effectiveness of internal process controls to mitigate risks. Most supply chain companies conduct systems and product audits but are moving to ISO 9001:2015 process and risk assessments.

SUPPLIER PARTNERING

You must never try to make all the money that's in a deal. Let the other fellow make some money too, because if you have a reputation for always making all the money, you won't have many deals.
J. Paul Getty

Customer-supply partnering is a common step to induce suppliers to start the SCRM journey or if they already have risk-controls in place to mature their risk processes. The goal of customer-supply partnering is to establish a mutually beneficial long-term alliance or relationship. In some partnering relationships, external suppliers are treated as an extension of the end-product manufacturer.

CUSTOMER-SUPPLY PARTNERING RISKS

Customer-supply partnering can take several forms. An end-product manufacturer may source to multiple suppliers. Or, an end-product manufacturer may single-source. Or, there are various options in between.

Single-source or multiple supplier sourcing both have risks. For example, Samsung's Galaxy Note Seven phone was removed from the market because its Lithium Ion batteries would overheat resulting in fire risk. Eventually, airline regulators outlawed their use on airplanes. Samsung permanently halted production of the smart phone after one month of their recall. Samsung is a global company with thousands of suppliers. Size risk from multiple suppliers played a key role in its RBPS and RBDM as the below quotation indicates:

> "A design flaw should have been caught during review and testing and this is much harder to do at global scale with multiple suppliers and factories for the same part" reported a VP at Forrester Research.[140]

Looking at the risk-benefit equation from the supplier's point of view, the supplier wants to be a single-source supplier and partner with the end-product manufacturer. The supplier may even share some risks with the end-product manufacturer. On the other hand, an existing supplier may not disclose the type of risks or the impact of its risks for fear of losing a contract.

HOW TO SECURE SUPPLIER ENGAGEMENT

An end-product manufacturer usually has two ways to get a supplier to adopt new practices: 1. Induce the supplier (offer rewards) or 2. Consequence the supplier (offer threats and punishment).

We are a great believer in the former – offering inducements. Suppliers are more apt to work with an end-product manufacturer if they are rewarded with demand, price, technical, manufacturing, and other inducements so the supplier can match its risk-control processes and capabilities with those of the end-product manufacturer. In these partnering relationships, key external suppliers are treated similarly as internal parts suppliers. External suppliers are monitored and expected to improve their risk-control processes at the same rate as internal suppliers. External suppliers are also expected to follow the end-product manufacturer's risk procedures, adopt similar risk-control processes, and establish similar communication protocols.

EVENT BASED SCRM INITIATIVE

While inducements are the optimum method to obtain supplier participation, downside risk or loss is often the initial driver for SCRM. Unfortunately, it is human nature to focus on current customer-supply practices until an event with a potential loss has occurred, then the SCRM initiative is started. The loss may entail financial, market share, reputation, or other types of consequences. For example, production loss may come in the following forms:

- **Complete loss of shipment.** Examples include cargo loss or earthquake loss.

- **Product damage.** Examples include product security, counterfeits, or expired products.

- **Shipment delay.** Examples include bad weather or shipment delays.[141]

VISUAL METAPHOR FOR PARTNERING

A visual is a powerful metaphor on how things work or how we hope they 'should' work. In customer-supply partnering, all the key players are on the same side-of-the-table, which is often based on openness and trust. Opposite side-of-the-table relationships imply self-serving and even adversarial relationships. Same side-of-the-table SCRM relationships are mutually rewarding, long-term, and trust based. Suppliers are induced not consequenced. Suppliers have the long-term win in mind.

Do same side-of-the-table relationships work? Yes. The auto and electronics sectors have developed long-term, close relationships with suppliers. Auto suppliers, in some cases, have been induced to locate their plants as close as possible to the end-product manufacturer's plant so their respective assembly processes are tightly integrated with frequent deliveries of small lots of parts for just-in-time assembly.

SUPPLY PARTNERING BENEFITS AND RISKS

The benefits of being a supplier to a large end-product manufacturer are evident. The supplier may secure large contracts, guaranteed margins, manufacturing technologies, special financing, and technical assistance.

To limit the risk exposure of a single-supplier not delivering, operations managers may also work with a prime and an acceptable alternate supplier. One supplier gets the lion's share of the business based on its risk-control processes. These two suppliers at the end of the year are objectively evaluated and business is allocated to each depending on past year's performance.

DON'T BOIL THE OCEAN

We come back to this point several times in this book. We have seen too many companies architect, design, deploy, and assure their SCRM initiatives without understanding and addressing their risk appetite. The answer to this question determines the breadth and depth of the SCRM initiative as well as investment.

Risk appetite is a critical SCRM discussion. How far should the end-product manufacturer go to 'appropriately' and 'reasonably' manage risks. The issues of 'appropriateness' and 'reasonableness' are moot since there is a point of diminishing return in the investment of SCRM. Some end-product manufacturers scope their SCRM at first-tier suppliers. Some go to second-tier suppliers. Or, the end-product manufacturer may stop its SCRM initiative with the first-tier supplier and require all first-tier suppliers be responsible for managing their supply risks. Or, the end-product manufacturer may want to trust but periodically audit lower tier suppliers.

SCRM SCOPING OPTIONS

An end-product manufacturer did not address its supply risk appetite and did not realize its SCRM ROI. Why? It tried to do too many SCRM initiatives at once and lost track of the benefits. In order not to not boil the ocean, an end-product manufacturer may want to initially limit SCRM program to first-tier suppliers or even only critical first-tier suppliers. Then the end-product manufacturer can request first-tier suppliers to maintain the same type and level of risk-controls on second-tier suppliers and down the value chain.

Lower tier suppliers can impact the end-product manufacturer disproportionately. A fourth-tier supplier may not conform to social responsibility practices, such as by hiring child labor or requiring forced labor to make products or provide services. The challenge is that manufacturers cannot audit or evaluate all lower tier suppliers. These are risks the end-product manufacturer simply needs to accept.

Some end-product manufacturers may decide to scope their SCRM at different levels such as:

- **First-tier supply level.** In this case, the end-product manufacturer may manage first-tier supplier risk and then require first-tier suppliers to manage their supply risks. This continues down the supply chain.

- **Second-tier supply level.** Some end-product manufacturers may assess their first and second-tier suppliers. SCRM stops with second-tier suppliers because the incremental investment to manage second-tier supply risks is not worth it. In other words, the end-product manufacturer decides to accept these risks.

- **Lower tier supply level.** The same RBPS and RBDM logic extends down the supplier chain. The end-product manufacturer is willing to accept lower tier supply risks. The end-product manufacturer may want to trust but verify and periodically audit lower tier suppliers based on perceived risks.

ISO 9001:2015 CERTIFICATION PROCESS

Practice what you preach.
Plautus, Roman Playwright

In this chapter, we use ISO 9001:2015 as an example of supply certification requirements for a candidate supplier based on an international standard. There are now supply certification standards for cyber security (ISO 27001), supply chain security (ISO 28001), and environment (ISO 14001). Also, there are supply certifications designed and tailored for industry segments such as aerospace (AS 9100) and automotive (IATF/TS 16949) sectors. In this chapter, we focus on the generic ISO 9001:2015 and IATF/TS 16949 as examples of supply certification approaches.

ISO 9001:2015 MANAGEMENT PRINCIPLES

ISO 9001, the international Quality Management System (QMS) standard, is the most widely used international standard. It was first developed in 1987 and has gone through 5 revisions.

ISO 9001 in the 1987, 1994, 2000, and 2008 revisions was often used as the basis of a quality based, customer-supply certification. It was a good to great framework to base the quality certification largely because of its eight management principles, specifically:

- **Customer focus.** End-product manufacturers meet final-customer requirements and attempt to exceed customer expectations through unique experiences.

- **Leadership.** Leaders establish the organizational direction and provide unity of purpose.

- **Involvement of people.** People are the core asset of the end-product manufacturer.

- **Process approach.** Results are achieved when activities, people, and resources are managed as an integrated process.

- **System approach to management.** Systems approach to management ensures that interrelated resources are identified, understood, and managed effectively and efficiently.

- **Continual innovation.** Continual innovation of all organizational elements is a permanent end-product manufacturer objective.

- **Factual approach to decision making (RBDM).** Effective and efficient decisions are based on data analysis and reliable information.

- **Mutually beneficial supplier relationships.** End-product manufacturers and their suppliers have interdependent processes that create stakeholder create.[142]

With the introduction of Risk Based Thinking (RBT) in ISO 9001:2015, the standard can now be used as a framework for risk based, customer-supply certification.

ISO 9001:2015 EXPANSION TO RISK
In the latest iteration of ISO 9001:2015, ISO standard developers clearly want risk to be part of a company's core business mission by expanding the boundaries of the Quality Management System, including:

- Conformance to the QMS requirements.

- Strategic and overarching view of the end-product manufacturer's competitive environment.

- Executive management involvement with the Quality Management System and Risk Based Thinking (RBT).

- Application and deployment of RBT in product development and the development of quality objectives

- Needs and expectations of interested parties who can impact product specifications and other requirements.

- Organizational mission and risk-control approach based on an analysis of opportunities (upside risks).

- Prioritizing risks in strategic planning, specifically providing leadership and commitment planning as well as a review of intended outcomes.

ISO 9001:2015 CERTIFICATION

Certification is a doable task for many companies if a team and project management approach is followed. Similar to managing any complex project, such as ISO 9001:2015 certification, executive management must understand and approve the project; resources are gathered and channeled to the appropriate locations and tasks; and supply stakeholders must see the benefits of the SCRM initiative.

The following are generalized steps for pursuing any standards-based supply certification like ISO 9001:2015 certification:

Step #1. Understand the ISO 9001:2015 certification environment. ISO 9001:2015 certification is a global phenomenon. An end-product manufacturer understands what competitors are doing, what customers want, and what regulatory authorities expect.

Step #2. Determine the benefits and challenges of ISO 9001:2015 certification. ISO 9001:2015 certification with RBT requirements is not easy. An end-product manufacturer pursuing ISO 9001:2015 certification or deploying quality systems should conduct a basic cost-benefit analysis of pursuing certification. ISO 9001:2015 is pursued to improve business operations, preferably not to get a 'ticket punched'. In the latter case, the end-product manufacturer may require ISO certification for a supplier to be on the approved bidders list. The supplier then develops quality and risk evidence such as documents to comply with the end-product manufacturer requirements.

Step #3. Secure management commitment. Executive management is fully committed and authorizes the necessary resources including people, monies, and equipment. Executive management must also actively participate in the certification by establishing project teams, maintaining organizational enthusiasm, providing direction, and reconciling differences. Middle management is committed because they provide the members for the ISO 9001:2015 project team.

Step #4. Plan for ISO 9001:2015 certification. ISO 9001:2015 certification is thoroughly planned so resources are not wasted or the project team does not go down dead ends.

Step #5. Organize for certification. Often, a multi-disciplinary project team tackles ISO 9001:2015 certification. The project team plans and develops management system documentation and evidence of compliance. The team also verifies that processes and systems are in place and operating properly. The team ensures that certification is implemented on time, under budget, and satisfies stakeholder requirements.

Step #6. Train, Train, Train. ISO 9001:2015 has new Quality Management System requirements, such as Risk Based Thinking (RBT), executive management commitment, new business processes to be implemented, risk audits to be conducted, corrective actions to be implemented, and procedures and work instructions to be written. All of these require training.

Step #7. Conduct the preassessment. The preassessment compares the 'what is' against the 'what shall/should be' as specified by the ISO 9001:2015. This is called ISO 9001:2015 requirements gap analysis.

Step #8. Flowchart processes and develop requisite evidence of compliance. Compliance is demonstrated through documentation or other forms of evidence. Developing documentation and mapping processes are the most time consuming and expensive elements of ISO 9001:2015 certification. While ISO 9001:2015 does not require formal documentation, evidence such as artifacts of compliance have to be developed. Three levels of ISO quality documentation are traditionally prepared, specifically quality manual policies, procedures, and work instructions.

Step #9. Select a certification body or registrar. Trust, service, and responsiveness are probably the critical factors for selecting a registrar or Certification Body. The registrar is an independent third-party that audits a company for compliance to ISO 9001:2015 criteria. This is largely a commodity service for which most registrars charge similar fees for similar services.

Step #10. Maintain certification. Once an end-product manufacturer secures certification, the company must actively maintain certification. ISO 9001:2015 certification is a long-term process of continuous auditing to ever-higher revised quality standards. The standards are revised every six to seven years incorporating higher quality and business process requirements. As well, the registrar conducts a surveillance audit of a company every six months or every year.[143]

AUTO INDUSTRY CERTIFICATION MODEL

No progress is going back.
Proverb

For years, U.S. companies dominated world markets by being vertically integrated. In **Purchasing Strategies for Total Quality**, I wrote,

> "Conventional wisdom said that if a company owned the sources of raw material, processed the raw material, designed the products, machined, fabricated, marketed, and finally distributed the products, products and market share would be assured. Vertical integration offered the advantages of standardization of products; control of operating, marketing, and distribution channels; and size and cost efficiencies."[144]

This could be seen in the auto industry where automakers controlled the entire design, manufacturing, distribution, and marketing chain. Little was purchased and almost everything related to developing an automobile was performed in-house. It was believed that if the end-product manufacturer owned the sources of raw materials, designed the product, and controlled the mechanisms for selling distributing, and servicing the product, the manufacturer could control risks, product delivery, quality, service, and costs.

CONTEXT: Mature SCRM Manufacturing Practices

- Predictive or preventive maintenance.
- JIT/continuous-flow production.
- Focused-factory production systems.
- Quick-changeover techniques.
- Bottleneck/constraint removal.
- Cellular manufacturing.
- Risk-control in most operational areas.
- Pull system/kanbans.
- Lot-size reductions.
- Competitive benchmarking.

Now, the automotive sector outsources much of its production and design to suppliers through sophisticated customer-supply certification models.

AUTO INDUSTRY CERTIFICATION MODEL

Over the last few years, high-tech companies, aerospace, and automakers entrusted suppliers with developing and providing larger assemblies of products. To reduce costs and improve delivery, automakers required first-tier suppliers to build entire systems, modules, or assemblies instead of just supplying parts and components. These suppliers were called system integrators.

Automakers required their suppliers to continuously improve and innovate by moving up a certification (maturity) curve such as the following:

- **Level 1: System is assembled by the supplier.** Component parts, cost responsibility, and product development responsibility remain with the end-product manufacturer.

- **Level 2: End-product manufacturer retains control of suppliers and sets cost.** End-product manufacturer does product development, but first-tier suppliers take responsibility for product production, quality assurance, and parts ordering.

- **Level 3: Full turnkey.** End-product manufacturer places full responsibility for an assembly with the supply-partner. The first-tier supplier assembles a system of products and has full sourcing and product development responsibilities for that assembly.[145]

IATF/TS 16949

ISO/TS 16949 is the auto industry equivalent of ISO 9001:2015, but has sector-specific requirements. ISO/TS 16949 was first developed in 1999 to combine ISO 9001 along with automotive sector-specific requirements to ensure quality throughout the automotive supply chain system.

IATF 16949:2016 may represent the future of supplier certifications because of the following attributes:

- Automotive Customer Specific Requirements (CSR):
  - Review of interested parties and their requirements shall be considered in the setting annual performance targets in the organization.
  - Supporting functions (on-site or remote) must be clearly defined in the scope of the Quality Management System (QMS).
  - Outsourced processes shall be included in the QMS.
  - Documented process for the management of product safety.
  - Special training for personnel involved in this process.
  - Identification of statutory and regulatory product-safety requirements.
  - Identification and control of product-safety characteristics.
  - Special approvals from organization's customers are necessary.
  - Transfer of product-safety requirements to sub-tier suppliers.

- Risks and preventive actions:
  - Continual risk analysis should include a minimum of potential and actual recalls, field complaints, scrap, and rework.
  - For manufacturing processes and infrastructure contingency plans shall be defined with periodical tests for effectiveness, review, and updates.
  - Contingency plans shall include notification to customers, restart of production and production stop procedure, also in cases where the procedure was not followed.

- Supplier selection process criteria:
    - o Specific requirements for suppliers of software with software quality insurance system.
    - o Supplier QMS to comply with IATF 16949 or periodic second-party audits are required.
    - o Definition of second-party audit process.

- Requirements for specific tools for process and product audits:
    - o Internal auditor qualifications include customer specific requirements from different Original Equipment Manufacturers (OEM's) and request to have more evidences of knowledge and training for core tools, customer requirements, etc.
    - o Maintenance of internal auditors' qualifications shall be defined (e.g. evaluation of audit results, execution of minimum number of audits per year, auditor development).

- Management input and review:
    - o Addressing risks.
    - o Process effectiveness and efficiency.
    - o Product conformance.
    - o Warranty conformance.
    - o Review of customer scorecards.
    - o Potential field failures through Failure Modes and Effects Analysis (FMEA).
    - o Management review output shall include an action plan when customer requirements are not met.

- Embedded software:
    - o Requirements included for Product and Process design.
    - o Requirements broken down to suppliers of the organization.
    - o Consideration of interaction of embedded software with the product in case of customer complaints.

CUSTOMER-SUPPLY PARTNERS

Automotive end-product manufacturers often prefer to develop turn-key partners for key assemblies such as for steering systems, computer controls, and engines. In these partnerships, there is often more trust with open innovation and shared Intellectual Property between the end-product manufacturer and key suppliers.

Supply chains in the automotive, telecommunications, and space sectors are often elaborate involving multi-tiered suppliers. In complex supply chains, commodity suppliers operate on a low-margin and high-volume business model, which often requires continuous production and monitoring. If there is a hiccup in the supply chain, this can lead to disruption with lower tier suppliers. These low-margin, high-volume suppliers often have not invested in increased production capacity so supply chain disruptions will have immediate impact on profitability.

SCRM CERTIFICATION BENEFITS

We have to make sure, at Apple, that we stay true to focus, laser focus – we know we can only do great things a few times, only on a few products.
Tim Cook, CEO of Apple

Most certification standards initially focused only on quality, cost, or delivery requirements. Now, supply certifications are more holistic and integrate risk.

BUSINESS CASE FOR CERTIFICATION

Too often, we have seen the rationale for standards based certification based on weak reasoning, such as 'we have been certified for 15 years,' 'we have always done it this way,' 'end-product manufacturers require certification,' and so on. SCRM supply certification may require a higher level of rationale and due diligence.

SCRM supply certifications offers benefits, including:

- End-product manufacturer benefits.

- Internal benefits.

- Customer-supply partnering benefits.

CONTEXT: Supply Chain Tips and Tools

- Plan and deploy inventory effectively.
- Provide predictable delivery performance.
- Create new products and services.
- Reduce order fulfillment cycle time.
- Reduce products in stock.
- Decrease manufacturing cycle times.
- Reduce transportation costs.
- Reduce final-customer returns.

End-product Manufacturer Benefits

Supply risk certification provides the following end-product manufacturer benefits:

- **Assists in developing products.** Time-to-market has become an important driver to SCRM success. If supply process risks are controlled, capable, documented, and improving, a new product can be designed and developed more quickly. Specifically, supply certification ensures that:

  o Supplier organizational structure and risk foundation exist for rapid product development.
  o Supplier has processes to facilitate product development.
  o Supplier core processes are proceduralized and standardized.
  o Supplier processes are stabilized, capable, and improving.

- **Provides access to markets.** There are several accepted certification standards including those tailored to the auto industry, telecommunications, and ISO 9001:2015 certification for other companies. End-product manufacturers want assurance of the supplier's quality and risk capabilities. Compliance or certification to these standards provides risk-assurance that management systems are in place. Thus, products from a Portland, Oregon manufacturer is consistent with those from a similar Pakistani manufacturer. And, ISO 9001:2015 products move transparently from one market to another.

- **Fulfills contract requirements.** ISO 9001:2015 certification or equivalent compliance to standards appears in more commercial and government contracts largely due to the influence of the auto industry adopting system standards.

- **Establishes promotional credibility.** Markets are crowded with competing products and services. Product positioning becomes critically important. 'ISO 9001 registered or certified' often appears in magazine ads and promotions extolling the virtues of the end-product manufacturer or its suppliers. As well, ISO 9001:2015 certification helps the company stand out and differentiate itself from the competition.

- **Conveys operational integrity and systems risk-assurance.** ISO 9001:2015 certification conveys to an end-product manufacturer that a supplier, that is four tiers removed, has established internal control systems that have been audited by an independent and objective third-party. In some countries, an end-product manufacturer can look up a supplier in an ISO registry and determine if the supplier is qualified to produce a product.

- **Facilitates SCRM planning.** SCRM focuses on where supplier risks may occur as well as the type of risks that may occur. The question often arises how far should an end-product manufacturer go to evaluate risks in the supply chain? The challenge is many second-tier suppliers may have risk issues that will be passed on to their first-tier suppliers. Second or lower level risk may not be recognized by the end-product manufacturer. The end-product manufacturer is simply too far removed from the root-cause problem.

Or, if the end-product manufacturer realizes there are problems, how does the end-product manufacturer know if these are simply business noise. RBPS and RBDM assessment tools discussed earlier can be used to answer these questions.

Internal Benefits
Most customer-supply certifications originally focused on quality. Now, they emphasize cost reduction, on time delivery, information technology, and risk-control applications. The Malcolm Baldrige Performance Award is

sometimes used as a customer-supply certification and includes business performance and innovation criteria beyond quality.

Internal certification benefits include:

- **Facilitates business and risk planning.** Supply risk certifications may require identifying stakeholder requirements and developing plans on how these will be satisfied. ISO 9001:2015, for example, requires a supplier to identify the end-product manufacturer's risk requirements and then integrate these into its planning processes.

- **Is used to transform organizations.** American business is undergoing rapid change. A critical ability of a senior manager is to create a sense of urgency, define the direction, and then lead the SCRM change. Change goes by different words and may involve SCRM, lean management, JIT management, or other initiatives. ISO 9001:2015 certification is the first milestone in the SCRM journey.

- **Provides a common approach and model for pursuing the SCRM journey.** Certification ensures that supply chain stakeholders speak a common risk language and follow consistent rules. ISO 9001:2015 provides an accepted and universal platform and approach to supply certification. Baldrige Performance Award also has integrated Enterprise Risk Management and can be used as a guideline for the SCRM journey.

- **Is used in many sectors and end-product manufacturers.** End-product manufacturers in different sectors can use certification to evaluate supplier risk maturity and improve supplier risk performance.

- **Encourages wide adoption of global standards.** Extensive use and versatility are critical assets of certification. For example, government agencies, non-profit companies, schools, service organizations, and other end-product manufacturers are adopting and certifying to ISO 9001:2015 and similar global standards.

- **Assists in establishing supply chain baselines.** As emphasized throughout the book, the supply chain is a process or more specifically a series of integrated processes. Certifications can establish a baseline for understanding and flowcharting core and risky processes.

CUSTOMER-SUPPLY PARTNERING BENEFITS
Customer-supply partnering benefits include:

- **Provides insights on organizational interrelationships.** Certifications often detail and identify requirements for interrelated systems and processes. These are the backbone of a supply chain. When the end-product manufacturer and suppliers can map these, they can identify areas of interaction, constraints, interrelationships, and key decision points. These can then be leaned and risk managed (RBDM).

- **Encourages supply chain focus.** Certifications often require policies, procedures, and work instructions are developed and followed. Developing the process and risk maps allows supply process stakeholders to clarify risk responsibilities, mission plans, risk-controls, and SCRM objectives.

- **Facilitates internal process control.** Management review, internal controls, auditing, and closed loop corrective action are the core elements of all certifications. They ensure that key processes are in control, are capable, and are improving. If deficiencies and risks arise, they are root-cause corrected. ISO 9001:2015 implementation also establishes the following internal control systems: design control, supply management control, and document control.

- **Assists employees in understanding and improving processes.** Properly implemented, certifications encourage supply chain stakeholder 'buy-in.' Stakeholders become engaged in identifying, mapping, proceduralizing, auditing, and correcting process nonconformances (production product defects).

- **Encourages self-assessment.** As mentioned, internal auditing is a major; some would say the most important element of most supply chain certifications. Internal auditing with the end-product manu-facturer and supplier results in the following benefits: processes are continuously monitored; if deficiencies occur, they are root-cause eliminated; and systemic and chronic problems are uncovered (RBPS).

- **Maintains internal consistency.** Process consistency is the hall-mark of high maturity and capable supply chains. Unknown or un-expected variation or risk, the opposite of consistency, is the nem-esis of all supply chain processes. Stabilized or risk-controlled pro-cesses imply delivery consistency, product uniformity, accurate de-mand planning, and lower overall risk.

- **Ensures internal processes are lean.** Many end-product manu-facturers are on autopilot. They do things because that is the way they have always been done. A certification process forces com-panies to ask: 'what are we doing, do our activities add value to the chain, and can things be done better'? These are all key lean ques-tions.

- **Ensures product development and design changes are con-trolled.** Certifications focus on upstream events as in preventing design nonconformances. Cycle time management and rapid prod-uct development are essential for the survival of many companies. Product flaws can result in an expensive recall. The solution is to fix problems at the source, for example in the design stage when the least value has been added to the product. A robust, low-cost, and high-quality product can also be designed at this stage. Cor-rective and preventive systems are established so deficient prod-ucts will not end up in the final-customer's hands. Formal design reviews also minimize after-the-fact design modifications and flaws (RBPS).

- **Encourages customer-supply problem solving (RBPS).** Proce-duralized operations reveal reengineering and innovation opportu-nities (RBPS).

CONTEXT: Auto Company Technology

Manufacturing companies have moved from analog to digital to robotic technology within a generation. This is now occurring with most if not all end-product manufacturers. Let us look at the auto industry.

User friendliness is a critical design feature in a vehicle and is now a critical value contributor and differentiator among auto manufacturers.

In the analog phase, automobile manufacturers used to worry about dash-board instrumentation with too many buttons, switches, and gages. The buttons were either push-pull, too small, too large, too close together, or too difficult to read. This could also be dangerous if a busy instrumentation cluster distracted the driver.

Then they moved to the digital world, where the instrumentation cluster was digital with cameras, warnings, sensing devices, and other driving distractions.

Now, each vehicle manufacturer is looking at autonomous or self-driving vehicles with much more electronics. These vehicles are mature vehicles with specialized electronics, software, and robotics. In a few years, these vehicles with be able to self-drive.

FINAL THOUGHTS ON SUPPLY CERTIFICATION

I say an hour lost at a bottleneck is an hour out of the entire system. I say an hour saved at a non-bottleneck is worthless. Bottlenecks govern both throughput and inventory.
Eliyahu M. Goldratt, The Goal

The conventional wisdom for ten years was it was more cost effective to offshore then to find domestic suppliers. Now this is changing the basic ideas of supply certification.

As well, other things have changed. More countries have domestic sourcing programs, focusing on 'Made in the USA', 'Made in India', and 'Made in China'. These issues add more risk to the 'make or buy' decision.

OFFSHORING VS. DOMESTIC SOURCING

The cost of offshoring has become prohibitive for more products. The driver of globalization and global sourcing was often cost reduction. Operating and supply chain business models were designed based on this assumption. The challenge is that many end-product manufacturers outsourced core processes not really understanding the many factors and risks of global sourcing.

Ten years later, low-cost is less of a driver for offshore sourcing. Why? End-product manufacturers would justify the benefit of outsourcing based on initial product or service costs. The full cost of offshore sourcing was often not captured, but with ten years' experience of offshoring, the true or full costs are now better understood. As well, the risks of global sourcing and extended supply chain are better understood along with possible points of failure and the full range of threats. More companies are now weighing the risks and rewards of global sourcing and are choosing domestic sourcing.

ALTERED SOURCING AND SUPPLY CHAINS

Shopping online by consumers is also shifting the sourcing paradigm. Online shopping is resulting in a functional reshaping of retail supply chains. We saw this recently as we shopped for eyeglass frames. The eyeglass company had a brick and mortar presence as well as a strong online presence. The brick and mortar shop had an optometrist offering eye prescriptions and knowledgeable sales people. The store also had a strong online catalog of eye frames. The store optometrist would test the customer's eyes and give the customer the prescription in the store, but would take the order from the online catalog for specialty, premium-priced frames.

Why is this occurring? Big box retailers have large amounts of unsold goods in inventory or on the shelf. The cost and quality of these products are high. Consumers purchasing these products are keeping them longer. Impulse buying is down. Consumers are more educated in what they want, willing to pay for, and quality of the product.

The results of the above changes result in uncertainty and risk for branded companies, end-product manufacturers, and their suppliers. Branded companies may face uncertain demand. They place smaller orders with their suppliers. Their selection is smaller than usual. The ultimate result: fewer people are employed.

As consumers change their purchasing habits, end-product manufacturers are searching for solutions using risk management. This is the future of retail. Target, J.C. Penney, and retailers are moving away from stores with specific merchandise for all final-customers. They are providing examples of merchandise on the shelves and are moving products to strategically located warehouses or even having products shipped directly from the supplier to the final-customer, like the eyeglasses we ordered.

INFORMATION IS POWER

In the New 2.0 Economy, core competencies and innovative information may evolve into the distinguishing competitive feature of the supply chain. In the e-commerce world, the supply chain is more often web based.

Many of today's supply chain gurus argue that ecommerce and Internet are about making communications powerful, instantaneous, and cheap. Product delivery, price, and quality are all available on the web through web based communications. A consumer or buyer can use a smart phone to check a competing price or track package deliveries in real-time. The buyer can comparison shop prices using Amazon or EBay. The buyer can use consumer ratings to check quality. The result is that communications has shifted purchasing power to the consumer.

Communications has also shifted information power to end-product manufacturers because they are accumulating critical final-customer and supply chain information. The companies that know the final-customer profile have a business advantage and the opportunity to make additional monies. The companies that have supply chain information can also optimize quality, costs, and delivery to assure high margins.

If an end-product manufacturer has a killer new idea, then it can increase economies of scale by sharing it with supply chain partners. The Internet and ecommerce help ensure that reliable, timely, and valuable information is available to everyone.[146]

SUPPLY DEVELOPMENT THOUGHTS

The best companies seem to have a culture that thrives on change technology, end-product manufacturer, system, competitor, and marketplace changes. The faster the rate of change, the more these end-product manufacturers thrive. It is a matter of anticipating final-customer requirements and leapfrogging the competition. These 'fast companies' also seem to form supply confederations more easily than others.

In today's economy, it is difficult to promote supply development and innovation when there have been internal layoffs and supply reductions. The surviving employees may grieve for their departed compatriots. Suppliers are worried about their own contracts and future. In this climate, a SCRM initiative may seem like a euphemism for supply reduction. Suppliers want assurance they will be part of the team next month.

What makes SCRM also difficult is that many supply chain details are unknown or deal with issues the operations manager cannot control, induce, or influence. In this atmosphere, the operations manager's role is even more challenging. As one operations manager once said to me, he could 'lead, follow, get out of the way, or be fired.'

SUMMARY

Supply chain management is all about measuring internal as well as supply performance. One-sided SCRM measures can cause problems. Trust is gained through mutual wins. Unfortunately, the reengineering and downsizing craze that hit the supply chain several years ago and continues today fractured much good will between end-product manufacturers and their suppliers. Many end-product manufacturers reduced their supply bases by 50% or more. As well, many suppliers became cynical as they discovered that many companies imposed supply requirements that end-product manufacturers did not follow themselves. Suppliers said that the customer-supply relationship was one sided. In other words, 'do as I say' not 'as I do'.

All global end-product manufacturers focus on and manage first and sometimes second-tier suppliers. However, risks in a fourth-tier supplier in a remote corner of the world can disrupt long supply chains. Terrorist attacks, natural disasters, diseases, earthquakes, plant closures, and labor disruptions can all impact the supply chain.

NEXT CHAPTER

In the next chapter, we discuss supply selection, another important element of SCRM deployment.

CHAPTER 11:
SCRM DEPLOYMENT –
SUPPLY SELECTION

WHAT IS THE KEY IDEA IN THIS CHAPTER?

According to a recent survey, 83% of end-product manufacturers have been impacted by supplier problems. Another interesting item from the survey was 66% of manufacturers expect supply risks and are working closely with suppliers to preempt them.[147]

Supplier selection is a critical element of SCRM Risk Based Problem Solving (RBPS) and Risk Based Decision Making (RBDM). In this chapter, we discuss how RBDM and RBPS can be used for supply selection.

SOURCING APPROACHES

If you can't describe what you are doing as a process, you don't know what you're doing.
W. Edwards Deming

A critical issue for many global companies is to determine whether to use a single-source, use multiple-sources, or have a single-source prime and a secondary alternative supplier. These options are various forms of risk hedging strategies. These risk decisions (RBDM) are based on criteria such as location, risk-controls, economies of scale, learning curve of the supplier, costs, use of proprietary engineered products, labor, distance to the assembly point, type of production planned, and many other factors.

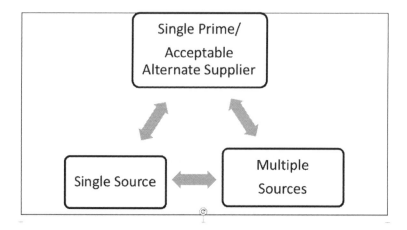

*Figure 33: **Sourcing Approaches***

THREE SOURCING APPROACHES

So in general, there are three common approaches to sourcing:

- **Single-sourcing.** Single-sourcing works with critical supply-part- ners or commodity suppliers. But, the risk is the supplier may be disrupted or curtail production for some reason such as a tsunami.

- **Multiple-sourcing.** Multiple-sourcing works with suppliers located in different parts of the globe and results in low to medium risks of product disruption. The risk is the supplier may not be motivated to work with the end-product manufacturer to provide requested prod- ucts or special services if the relationship is based on low-cost.

- **Single prime and acceptable alternative supplier.** Strategy is used with critical suppliers where supply chain business continuity is critical. The risk is the prime or acceptable alternate supplier may rest on its laurels and not adopt new SCRM requirements.

The single prime and acceptable alternate supplier business model is a hybrid of the previous two. In this option, there is now a balance between having back up materials in stock for critical parts and maintaining lean production of non-essential parts.

SOURCING FACTORS

Many supply chain risk-control decisions can involve determining the acceptable level of risk from a make/buy decision, finding a new supplier, or deciding on a 'sole-source' supplier. Let us look at the risk of choosing a 'sole-source' supplier compared to 'multiple-sources.'

The risk of having a 'sole-source' supplier is balanced against having multiple suppliers. While it is easier to manage one supplier, this decision entails higher risks. There is a possibility of an act of god, such as tornado or a strike could impact the sole-source supplier.   What are the consequences of a Black Swan event occurring (RBDM)?

On the other hand, multiple suppliers must be managed more carefully than a single-supplier. Multiple or offshore suppliers require more effort, resources, and time to manage. So, the risk of a single-source supplier is balanced against the cost, time, and effort of managing multiple suppliers.

SINGLE-SOURCING

Ten or fifteen years ago, lean manufacturing and just–in-time delivery were the core tenets of supply management excellence. These concepts were seldom challenged and were the fundamentals of operational excellence. However with extended supply chains, single-sourcing amplifies supply disruption.

Coupled with this, we are seeing more domestic manufacturing or domestic sourcing. This has profound impact on the distribution sector. Shipping ports have idle equipment. Ships and trucks are not used. Shipping professionals are not employed or underemployed. Capital expansion is delayed or simply canceled.[148]

However, there are significant risks with single-sourcing. Think of putting all of your eggs in one basket and the basket is dropped. This is what happens with a supply chain disruption. Some sectors with sensitive products offer higher risks that should be identified and preemptively controlled or hedged.

CONTEXT: Supply Partnering Benefits and Challenges

Upside:
- Internal risk-controls.
- Supplier risk-controls.
- High volumes.
- High recognition.
- Preferred access to future projects.
- Improved manufacturing processes.
- Quality gains and lower costs.
- Access to specialized designs.
- Access to production expertise.
- Lower costs.
- Ability to synchronize just-in-time operations.
- Quicker turnaround time.
- Fewer problems in design and product change-overs.

Downside:
- Minute scrutinization and operational pressures.
- End-product manufacturer presence.
- ISO 9001:2015 or other certification requirements.
- End-product manufacturer required controls.
- Annual 'report cards'.
- Order levels tied to performance.
- Ever tighter requirements.
- Vulnerability to sharp demand swings.
- Possible loss of other end-product manufacturers.
- Evaluation of past history.
- Assessing security of information.
- Loss of sensitive or proprietary information.
- Loss of end-product manufacturer confidence if failure occurs.
- High opportunity costs.
- Loss of market share.
- Loss of credibility among stakeholders and competitors.

DEPARTMENT OF DEFENSE SINGLE-SOURCING REQUIREMENTS

While single-sourcing is more economic and efficient, the risks for critical equipment and services may not be acceptable to a government agency or end-product manufacturer.

For example, the U.S. Department of Defense (D.O.D.) requires large defense contractors to report on their single-sourcing, specifically:

- Include a list of critical components provided by single-source suppliers.

- Identify major defense acquisition programs with operational implications of single-sourcing.

- Identify risk management actions with associated deployment plans and timelines that D.O.D. can take to prevent negative operational impact in the event of a loss of such suppliers.

- Identify severity of operational impact of the loss of such suppliers.[149]

TOYOTA CHALLENGE

An end-product manufacturer such as Toyota will attempt to find the right balance between risk, economics, and efficiency with suppliers. The right balance between the risk of a possible disruption and the rewards of single-source partnering is a critical risk-control decision. If Toyota decides to single-source, a supplier disruption could halt auto assembly. If Toyota decides to multiple-source, then this would add additional costs. This is a fine balance (RBDM).

Along with this balance, SCRM involves continuous improvement and cost cutting. Some end-product manufacturers are asking suppliers for 2% discounts on a year-by-year basis. This forces suppliers to continually redesign products or manage their own suppliers more carefully. This is another new normal of SCRM.

SUPPLIER SELECTION PROCESS

Supplier selection and segmentation has even been called the 'next best practice' in supply management. [150]
Stuart Ian

Best-in-class supply chains select and develop suppliers carefully. They may use far fewer suppliers than their competitors and spend less on materials than their industry rivals.

RESEARCH CRITICAL SCRM ISSUES

Identification of suppliers to meet end-product manufacturer requirements is a critical element in the selection process. The selection of suppliers is based on real and perceived risk-controls. For example, the selection of a commodity supplier would be based on product quality, schedule, and cost value attributes. The selection of a more critical supplier with a critical engineered product would be based on process capabilities, risk management capabilities, and shared-risk vision.

COMPILE SUPPLY DATA

The next step is to identify prospective suppliers that can manufacture the required products or provide the services. Suppliers can either be domestic or foreign. A domestic-supplier list is generated from personal knowledge, web search industry contacts, catalogs, trade publications, advertisements, or directories. The foreign supplier list is generated from chambers of commerce, departments of economic development, U.S. State Department, U.S. Commerce Department, trade representatives, foreign consulates, and trade publications.

Data from these sources is then collected to develop an overview of candidate suppliers. While many U.S. companies prefer to buy American products, a global economy and the need for competitiveness has forced companies to go overseas to obtain the best products at the most competitive price.

It is critical to obtain a holistic, 360^0 view of critical supplier operations and their suppliers to understand where the risks may reside. Risks can hide in spend data (savings leakage), waste (non-lean operations), and supplier risk data (non-compliance, reputational risk, viability, etc.).

BUILD REGISTER OF SUPPLIER RISKS

Evaluate candidate suppliers and narrow the list to those acceptable to the end-product manufacturer. This is the first cut. Approval depends on factors, such as quality, cost, delivery, technology, service, risk, and value factors.

It is critical to arrange preliminary supplier risks in a spread sheet and score them based on a heat map grid. Risks can be identified in terms of consequence and likelihood such as High-High (H-H), Medium-Medium (M-M), and Low-Low (L-L). The risk assessment at this stage does not have to be elaborate. It simply provides a risk overview or profile of the supplier.

Suppliers with High-High evaluations may be rejected based on their overall risk. The purpose is to segment suppliers to those that pose little or no risk to the end-product manufacturer. This process is done with existing and potential suppliers within the category.

Each candidate supplier may get a second look. The process is meant to be flexible.   So if there is sufficient rationale, then a second SCRM team can conduct another supplier review to ensure consistency and validation. The first time this is conducted, there will be little consistency. However, as each new supplier is added then the team will be able to get a list or register of supplier risk-controls.

Several risk-control related questions to consider at this stage include:

- How critical is the part or product?

- Is the part a commodity, high-tech, or proprietary product?

- Is there a long lead-time?

- How long does it take to select and qualify the supplier?

- Are there alternate suppliers for the product or service?

- Does the supplier add proprietary, confidential, or Intellectual Property to the product or service?

- What is the likelihood and consequence that risks may appear with a supplier over the next year or three years?

- What is the attitude of the supplier to risk-control the symptom and root-cause correct the problem (RBPS)?

NARROW SUPPLIER LIST FURTHER

The next step is to narrow the supplier list to those capable of manufacturing a product or delivering a service that conforms to specifications and poses the lowest level of acceptable risk to the end-product manufacturer.

Suppliers posing little or acceptable risk are usually considered 'approved'. The risk score becomes the baseline for the supplier selection and certification process. Some companies call this the inherent supplier risk score because the score reflects the supplier's risk with its present controls in place. No risk mitigation or risk-controls have been deployed with the supplier.

At this phase, a risk score based on the heat map is developed for the supplier's operational areas, projects, regulatory, reputational, cyber security, product attributes and financial areas. Scores are developed for cost, business, technology, quality, and risk at a low product level and then these are bundled and rolled up for a total risk score for the supplier.

One risk measure often used in manufacturing is process capability, which refers to the supplier's ability to control a process and consistently meet specifications. Capability implies the supplier has the technical and manufacturing competence to satisfy end-product manufacturer requirements. Capability is verified through Statistical Process Control, interviews, inspection of product characteristics, review of past manufacturing performance, audit of manufacturing processes, or exchange of product samples.

To evaluate capability, the supplier must have drawings, specifications, standards, and all the relevant information to make a conforming product.

Otherwise, the supplier will manufacture a product but not to the quality, performance, or reliability levels required by the end-product manufacturer.

SELECT PRIME SUPPLIER AND ALTERNATE

It is much easier to compare suppliers with similar products and services to determine a level of inherent risk or what is acceptable to the end-product manufacturer. The inherent risk is based on the supplier's ability to control risks within the risk appetite of the end-product manufacturer.

At this point, the end-product manufacturer must determine if the supplier: A. Is acceptable and put on the approved bidder's list; B. Is marginal and needs additional correction, risk-control, etc.; C. Requires significant improvement, or D. Is unacceptable and is removed from the list of suppliers. If the supplier is marginal, then a corrective action program may be initiated involving additional risk mitigation or risk-controls. Residual risk score at this point is shared with the supplier. The supplier is requested to develop additional plans for improvement or to mitigate risks in identified areas.

The last step is to select the prime supplier and maybe an acceptable alternate supplier. As opposed to single-sourcing, two sources are sometimes preferred to reduce the risk of receiving unacceptable products and to help ensure future price competition. SCRM coordinates this effort with quality, engineering, manufacturing, and other stakeholders. The final decision of selecting the prime supplier rests with senior supply management, because they are responsible if contract requirements are not met.

Formal evaluation plans are useful for making the supplier aware of end-product manufacturer requirements and expectations. Details for the evaluation are outlined and then attached to the purchase order (RBDM).

MEASURE AND MONITOR

It is valuable to have local teams in supplier locations who monitor suppliers; understand local customs and laws; and incorporate this knowledge into the supplier contract (and processes) to protect the end-product manufacturer from running afoul of laws and protecting their reputations.

CONTEXT: Key Questions For Selecting Suppliers

- **Is it critical to the end-product manufacturer?** What is critical to the end-product manufacturer, may not be on the supplier's radar screen.
- **Is it manageable?** If the end-product manufacturer demands SCRM, RBT, RBDM, RBPS, Six Sigma, lean, and Enterprise Risk Management (ERM) from suppliers, does the end-product manufacturer have the time to work with suppliers to monitor and advise them if they have immature processes?
- **Is it measurable?** There are a lot of low hanging fruit with quick returns in supply selection and development. Focus on these. The Pareto Principle in supply management says that 20% of the suppliers create 80% of the problems. So, start with the 'critical few' supply risks not the 'trivial many.'
- **Can it be done quickly?** There is low hanging fruit, but the end-product manufacturer may need a chain saw to cut the fruit down.

This brings risk cataloging full circle. It applies a scientific and systems approach to SCRM that asserts that risks are not static, but dynamic. Scoring, tracking, and proactively managing risks (ideally avoiding them altogether), and then applying lessons learned to the risk management framework is the preferred SCRM approach, rather than leaving things to chance and hoping for the best.

It is often difficult to develop a holistic view of an organization's supply chain structure. It becomes difficult to assess an organization's risk exposure with first-tier suppliers and even more difficult with second and lower tier suppliers. It also becomes difficult to make RBPS and RBDM based on poor quality data.

The end-product manufacturer needs to understand, correct, and prevent risks from occurring and recurring. However, most end-product manufacturers are not managing supply risks. The numbers are stunning! The reality is two out of five companies do not analyze the root-causes of supply chain disruptions. Another interesting fact from the survey reports that most of these companies do not have visibility of their supply chains, potentially resulting in future risks.[151]

CONTEXT: SCRM Supply Selection Tips

- Determine the end-product manufacturer's business model, which cascades to a SCRM business model.
- Establish the SCRM objectives that are aligned with the end-product manufacturer's business model.
- Determine supplier selection value attributes based on the end-product manufacturer overall SCRM business model and objectives.
- Identify which products and services can be internally sourced and outsourced.
- Establish selection value attributes for critical and non-critical suppliers.
- Develop a list of possible suppliers based on each product or service category.
- Categorize and prioritize suppliers based upon selection value attributes such as using a balanced scorecard with multiple criteria.

SELF MANAGEMENT IS KEY

SCRM looks to suppliers to risk-control and seamlessly integrate their processes into the end-product manufacturer's supply chain. In other words, suppliers are expected to self-manage their risk, quality, delivery, and cost processes. This implies there is no incoming material inspection, products are delivered just-in-time to the proper location, and overall contract cost reductions are shared with the end-product manufacturer.

So, brand owners and end-product manufacturers increasingly audit suppliers, certify them, and expect zero-defect level performance. If suppliers comply, rewards are shared and they move up the process maturity and capability curve. Suppliers are also induced by larger and longer-term contracts, technical assistance, and special equipment.

SHARED RISK – SHARED REWARD IN HEALTHCARE

Shared risk and return between end-product manufacturer and suppliers is a common practice in many sectors. Healthcare is an example. Medical end-product manufacturers used to charge a premium for implantable devices, especially if they involved new technologies. Now, large healthcare systems have data on the effectiveness and longevity of the implantable devices and can evaluate device performance across manufacturers.

Large healthcare systems are negotiating performance-based contracts where medical end-product manufacturers assume the financial risk of their implants. "Under those agreements, the device company is liable if its product Tyrx, an antibacterial sleeve, fails to protect against infections in patients receiving cardiac implants."[152] Risk sharing agreements are also occurring with pharmaceutical end-product manufacturers, ambulatory surgery centers, and service providers.

SUPPLIER SELECTION VALUE ATTRIBUTES

Without data, you're just another person with an opinion.
W. Edwards Deming

Most supply selection uses a 'weighted average method' to evaluate delivery, cost, quality, technology, and other performance factors. Each end-product manufacturer factor is given a relative weight according to its importance. For example, quality for high-value products would be rated higher than a commodity where cost would probably have a higher rating.

SUPPLY SELECTION VALUE ATTRIBUTES
Additional selection value attributes may include:

- Supplier performance.

- Location.

- Financial stability.

- Motivation.

Supplier Performance
Supplier performance history reveals if the supplier has been responsive in complying with contractual commitments. If a supplier has a history of satisfactory performance, this provides a degree of risk-assurance that material will be supplied just-in-time and at the specified quality levels. If the supplier is new, then a reference check of previous end-product manufacturers is essential.

Industrial sourcing decisions, which historically were made based on product price, availability, and service are also changing. If an operations manager was interested in buying a piece of industrial equipment, he or she first looked at the lowest price when evaluating competitive bids. Sometimes product availability was important if the piece of equipment was needed quickly. Service was important if the product was subjected to abuse, prone to failure, or designed to run continuously. Nevertheless, price was the primary consideration when buying a product (RBDM).

This situation has now changed. The primary factor influencing the purchase decision consists of risk related, value attributes while other factors are less prominent depending on the type of sourced products or services (RBDM).

Location
Location is important because logistical, plant location, just-in-time delivery, or technical assistance requirements mandate the supplier is located near the end-product manufacturer, not half way around the world. If problems arise, they are easier to resolve if the supplier is located down the street (RBPS).

Financial Stability
Financial stability indicates the supplier has been a going-concern for several years. This implies continuity and stability. An end-product manufacturer does not want a supplier of critical components to go bankrupt. A wayward second-tier supplier providing a critical component can disrupt contractual commitments of the prime contractor and can impact the entire supply chain.

Motivation
Supplier motivation is the supplier's attitude to work with the end-product manufacturer to resolve disputes in a timely and mutually beneficial manner. Disputes in customer-supply relations always arise. If the supplier is motivated, these differences are resolved quickly and amicably.

CONTEXT: Only As Good as the Weakest Link

The supply risk process chain is only as good as its weakest link or weakest supplier. So, supply development emphasizes:

- Core processes are flowcharted.
- Value creation and waste reduction are understood and pursued by critical supply stakeholders.
- Critical supply chain processes are stable.
- Critical supply chain processes are capable and improving.
- Supply chain choke points are identified and eliminated.
- Risk points are identified with sufficient controls to minimize risk.
- Information flows quickly to all supply chain stakeholders.

CAPTIVE SUPPLIERS

Another SCRM tactic is to secure captive suppliers. In other words, the end-product manufacturer may or may not have a sole-source arrangement with the supplier. But, the supplier has a sole producer arrangement with the end-product manufacturer. The supplier just provides products to the end-product manufacturer.

This SCRM tactic works for both parties. The end-product manufacturer gets a reliable source for critical products, at a known price, with known technology, with confidentiality, and a full understanding of needs. The end-product manufacturer may require high-volume suppliers to modify management and operational processes to comply with its risk-control requirements. A large end-product manufacturer may require suppliers to invest in additional property, plant, and equipment to increase or to improve production capacity to ensure supply of critical materials if there is a spike in demand.

There are several reasons why this is done. An end-product manufacturer may closely monitor its supplier performance and tie future order levels to risk, quality, cost, and delivery targets. The supplier gets a long-term contract with known margins. The supplier may also get advice and technical assistance from the end-product manufacturer. This arrangement is probably one of the best ways for an end-product manufacturer to integrate a

---

CONTEXT: Minimizing Supply Chain Logistics Risk

- Identify and assess current areas and levels of supply risk-control.
- Identify supply and delivery supply alternatives.
- Empower trading partners.
- Select vendors in different geographic regions who supply through secondary ports.
- Fully engage in supplier relationships.
- Take control of logistics processes.
- Jointly plan and collaborate about potential supply chain disruptions.
- Build flexibility into processes so the end-product manufacturer can promptly adapt to changes with minimum impact.
- Optimize inventory buffers and safety stock levels.
- Be proactive. [153]

---

supplier into the value chain. The end-product manufacturer can apply its process and information capabilities directly with the supplier.

SUPPLIER SELECTION RISK FACTORS

Knowledge is power.
Proverb

RBT is a journey much like we have seen in purchasing maturing to supply management and SCRM. The SCRM journey starts with RBT and as it matures, organizational value is enhanced through risk assessment, risk management, and ERM.

ASKING THE RIGHT RISK QUESTIONS

SCRM can often be boiled down to asking the right questions at the right time. One critical question to ask is 'what if'? The question helps the end-product manufacturer visualize potential risks in the supply chain and be able to quantify and prioritize them.

And often, there is no one right solution. The SCRM RBDM process is incremental based upon risk assessments at critical points of the customer-supply journey. So, let us look at some critical SCRM risk based, decisions that impact today's supply chains.

SUPPLIER SELECTION

There are several enterprise level SCRM tools. Supplier performance is evaluated based upon multiple value attributes that were often determined in the supplier selection process. Supplier selection can be based upon total cost, delivery, business model, process capability, quality metrics, engineering capability, and other key performance indicators.

Classification of supplier risks is an iterative process in most companies. Unless an end-product manufacturer has a history with dealing with a supplier, supply risks are often unknown and even unknowable. Why? Suppliers do not want to disclose confidential or proprietary information.

Also, a supplier will not disclose the type of risks or the impact of risks for fear of losing a contract. New supplier risks may have to be accepted by the end-product manufacturer or another way of saying this is that the end-product manufacturer assumes the supplier risk.

Several factors need to be considered in architecting, designing, deploying, and assuring a comprehensive risk management model for the supply chain. It usually starts at the top of the end-product manufacturer or what we call the enterprise level. This is the reason we always start with an enterprise approach to SCRM.

SUPPLY RISK DEVELOPMENT AS CORE PROCESS

Large end-product manufacturers prefer to outsource products to 'world-class' suppliers. What does 'world-class' mean? These suppliers have mature and capable processes to deliver world-class products on time and to the right location.

Supply development thus becomes the basis of supply integration. Partnering allows an end-product manufacturer to bring key suppliers into the product development process to take advantage of the supplier's core skills, knowledge, tooling and service.

Customer-supply partnering, joint ventures or strategic alliances are also preferred methods to penetrate new markets, provide a local presence, enhance product quality, lower risk exposures, create a market for new products, establish a local distribution network, resell products, integrate suppliers, and share process innovations.

SUPPLY RISK CRITERIA

SCRM often requires a new lens by which to review and evaluate risk factors associated with suppliers. Many end-product manufacturers still look at suppliers in terms of ISO 9001:2015 compliance, lean, Six Sigma, and quality capabilities. However, many global manufacturers have moved beyond looking at suppliers transactionally or in terms of product specification, or compliance, but are assessing each supplier in terms of its value contribution and risk to the entire value chain.

SUMMARY

Ten or fifteen years ago, supply selection was a straight forward process. Conventional supply management wisdom of the last ten years is now challenged, such as the benefits of single-sourcing, which can amplify supply disruptions. As part of SCRM, end-product manufacturers now challenge the conventional wisdom of single-sourcing and review alternatives to hedge against a supplier not delivering quality products or services.

SCRM changes can be seen in healthcare, where it is often called 'value based healthcare.' Healthcare is moving to risk-sharing agreements between providers, hospitals, and suppliers including pharmaceutical, medical device, and equipment manufacturers. About 83% of hospitals indicate that their supply chain management teams will be developing SCRM and risk based contracting over the next 3 years. [154]

NEXT CHAPTER

In the next chapter, we discuss SCRM risk-assurance and monitoring.

CHAPTER 12:
SCRM RISK-ASSURANCE

WHAT IS THE KEY IDEA IN THIS CHAPTER?
Supply monitoring and review are critical to all SCRM decision making (RBDM). Supplier monitoring provides the requisite level of risk-assurance to the end-product manufacturer based on its risk appetite and tolerance. Key performance metrics provide the foundation for the customer-supply relationship and development (RBDM).

MONITORING SUPPLIERS
Trust, but verify.
President Ronald Reagan

North American end-product manufacturers have established partnering arrangements with their suppliers to improve product and service performance. Traditional customer-supply relationships were based on anecdotal evidence to draw conclusions about supplier quality, cost, delivery, risk, and technology. The trend now is to rely on real-time objective, verifiable, and measurable information upon which to base supply chain partnering and purchase decisions (RBDM).

SUPPLY CHAIN FRICTION AND RISK
Friction is always present SCRM. End-product manufacturers seem to be shopping for lower prices. New technology with new applications seems to be changing the relationship between the end-product manufacturer and its suppliers. Competition and collaboration seem to be conflated. New compliance requirements seem to be increasing all the time. There is increased tension and risks between all parties in the supply chain. Thus,

customer-supply monitoring and risk-assurance have assumed more importance.

Traditional supplier monitoring involved some form of compliance, such as ISO 9001 certification, incoming material inspection, supplier quality auditing, and physical product verification. This normally focused on product conformance and standards compliance. The problem is that compliance is a yes/no decision. SCRM is moving to risk-assurance, which consists of more advanced supplier assurance and monitoring such as forensic reviews, control environment evaluation, risk mitigation, internal control, and other types of supply reviews (RBPS/RBDM).

SCRM systems monitoring and risk-assurance can become sophisticated. The end-product manufacturer may identify critical risk-control points in the supply chain; monitor anomalies, emerging variation, early warning signals of impending Black Swan events; and then develop risk-control response or treatment mechanisms. As well, extended supply chains are segmented into control areas where risk monitoring and control monitoring can be tied to value at risk.

RISK-ASSURANCE

Risk-assurance must be architected and designed to the end-product manufacturer's risk appetite and risk tolerance. Properly designed, risk-assurance can:

- Ensure SCRM integrates and deploys RBT, RBPS, and RBDM into all critical SCRM processes.

- Ensure the end-product manufacturer integrates and deploys proactive, preventive, predictive, and preemptive (4P's) management.

- Lead to peer, manager, and executive management reviews, and robust escalation of issues thus alleviating potential SCRM risks and concerns. SCRM executives can use information to monitor key SCRM project variances and exceptions, then explore and analyze the information to shed light into the exceptions and look for hidden risk trends.

- Monitor SCRM critical business processes using risk performance metrics to trigger alerts when SCRM problems arise. The alerts are used for intervention, correction, prevention, and ultimately improve SCRM processes (RBPS).

- Focus attention and risk-controls in the areas being monitored and measured. As the expression implies: 'what gets measured, gets done.

CUSTOMER-SUPPLY MONITORING

The big things you can see with one eye closed. But keep both eyes wide open for the little things. Little things mark the great dividing line between success and failure.
Jacob Braude

Auditing (monitoring), corrective action, and preventive action are probably the three most critical risk-assurance processes for monitoring supply performance. They 'close the loop' with SCRM problem solving (RBPS) and decision making (RBDM). They form the basis for intervening, correcting, and improving customer-supply processes (RBPS).

Supplier performance is evaluated based upon multiple value attributes that are often determined in the supplier selection process. Supplier selection can be based upon key performance and risk indicators such as total cost, delivery, business model, process capability, quality metrics, engineering capability, and process capability.

ISO 9001:2015 AUDITING
Quality auditing is part of most supply maturity models, customer-supply partnering relationships, and certification standards such as ISO 9001:2015. Suppliers are often required to designate a group of internal process, quality, and risk auditors.

For example, the objective of internal quality auditing is to determine the effectiveness of the Quality Management System and identify opportunities for innovation. ISO 9001:2015 states that:

"The organization shall plan, establish, implement and maintain an audit programme (s) including the frequency, methods, responsibilities, planning requirements and reporting, which shall take into consideration the importance of the processes concerned, changes affecting the organization, and the results of previous audits."[155]

Auditing can also be used for supply monitoring. Supply performance monitoring can review quality, delivery, cost, technology, and product variances. It can look at operational and quality effectiveness and efficiency. Or, it can look at product variances from a standard or engineering drawing.

Each of these variances from a performance target is a risk precursor or an early indicator of risk. For example, an increase or decrease in a value attribute (such as a dimension) from a target may indicate a nonconformance (product defect) or a lack of process control. Delivery variances either too early or too late would indicate shippers are not managing their schedules. Or, technology variances may indicate counterfeit parts or mislabeling. Each of these would be investigated to determine the root-cause of the deficiency (risk) and eliminate it.

CONTINUOUS SUPPLY CHAIN REVIEW

Customer-supply audits are one of the most effective monitoring and innovation techniques. Often, these assessments are called quality audits but also involve analysis of cost, technology, delivery, service, risk, and other factors. These audits are an official examination of supply chain processes, products, or people to verify process innovation, effectiveness, and compliance to policies, procedures, work instructions, and specifications. Audits can be conducted on a periodic or random basis and are performed unannounced or upon request.

Auditing is also critical to all supply chains to ensure the right products are delivered to the right location on time, in the right sequence, properly protected. In other words, audits assure key supply chain processes are running smoothly, are controlled, are capable, and improving.

Once the audit is conducted, audit findings are documented in a report that is circulated to authorized clients. The audit report may recommend preventive action, corrective action or no action. Usually, if action is indicated, a corrective action request (CAR) is written to initiate further action.

ADVANTAGES OF CUSTOMER-SUPPLY AUDITING

The following are advantages of conducting customer-supply audits:

- Provide independent and objective advice to management on the effectiveness, efficiency, and economics of risk-controls.

- Provide management with risk-assurance.

- Monitor customer-supply processes and products.

- Provide corrective action follow up.

- Identify areas of process innovation.

- Measure effectiveness of people, processes, product, and organizational innovation.[156]

AUDIT PROCESS STEPS

Customer-supply, internal quality, certification, and risk audits follow a systematic process of:

- **Step 1:** Preparation.

- **Step 2:** Preliminary conference.

- **Step 3:** Data acquisition.

- **Step 4:** Data interpretation.

- **Step 5:** Closing conference.

- **Step 6:** Feedback and action.

- **Step 7:** Root-cause risk elimination.

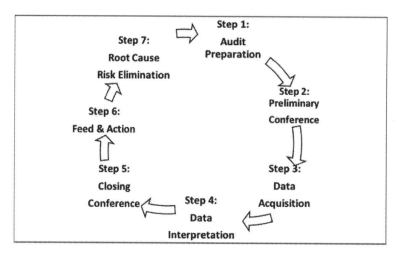

Figure 34: Audit Lifecycle

Step 1: Preparation

The auditee (the end-product manufacturer or supplier) is first notified of the impending audit. The auditee is advised when it will be conducted, who and how many people will be involved, and why it is conducted.

The audit team obtains background information on the auditee. Background information is obtained from financial statements, previous audits, and trade journal articles. For specific information, the team evaluates risks including the contract, quality manual, specifications, bills of material, drawings, layouts, flowcharts, reliability test reports, and customer service history.

Step 2: Preliminary Conference

The preliminary conference at the auditee's facility establishes audit ground rules. Any auditee fears are dispelled in this conference.

Step 3: Data Acquisition

The audit team or auditor collects program, process, people, and product information. The team interviews management and employees. The team reviews quality, policies, quality procedures, product plans, cost, technical capabilities, and specifications.

Step 4: Data Interpretation
The team interprets evidence and data collected during the audit. This is time consuming and labor intensive. The team understands the client's industry, internal processes, and products. With this knowledge, the team attempts to identify unusual variations, risk-control effectiveness, constraints, or waste and may recommend corrective or preventive actions.

Step 5: Closing Conference
The closing conference reviews preliminary impressions, problems, conclusions, and recommendations. If there are substantive areas of disagreement, these are appealed to the end-product manufacturer's executive management.

Step 6: Feedback and Action
Finally, the auditee is formally notified in writing of audit results. If deficiencies or nonconformances (risks) are found, the auditee (the supplier) is asked to correct them prior to approval of shipment of products. If the auditee does not agree with the recommendations and conclusions, then customer and auditee (supplier) management may resolve the differences.

If the supplier is evaluated for the first time, the audit team may approve the supplier, reject the supplier, or recommend further evaluation. If the supplier has been supplying products for some time, the audit is used as a monitoring tool to encourage continuous innovation and risk-control application.

Step 7: Root-cause Risk Elimination
Root-cause risk analysis investigates the cause and the chain of events that resulted in the problem. Each critical process step is assessed, possible causes are identified and examined until a root-cause is determined. Once this process becomes institutionalized, SCRM problem solving (RBPS) becomes a normal course of business. The challenge is a problem may have multiple causes, which make it difficult to determine the prime cause in a supply chain consisting of many sub-tier suppliers (RBPS).

The operations, supply, risk, or quality manager in an audit does not take the role of process owner. The process owner is the supplier who is still responsible for the process solution or project intervention. The operations manager can advise but the owner must decide on and implement the solution.

AUDIT DATA ANALYSIS

Audit analysis can reveal risks, variances, nonconformances (defects), and opportunities for innovation, as indicated by the following:

- **Risk analysis.** Areas of high supply risks and weak internal controls to minimize or eliminate the identified risks.

- **Cost analysis.** End-product manufacturer process, product, customer expectations, and operational cost variance analysis may indicate opportunities for cost reduction.

- **Type of and number of risk and/or deficiencies (nonconformances).** Deficiencies may include a final-customer complaint, cost variance, ISO 9001:2015 noncompliance, or product field failure. This data may indicate if the problem is isolated or if it is an early warning of a widespread problem (RBPS).

- **Location of deficiency.** The location of the deficiency reveals if a problem is isolated to a specific location, gage, or application or is a chronic widespread problem. This data when collated or analyzed can provide more clues for isolating the problem and determining its root-cause (RBPS).

- **Corrective action analysis.** Corrective action analysis determines if actions to eliminate the cause of the deficiency were successful (RBPS).

- **Preventive action analysis.** Preventive action analysis is a proactive analysis of data such as supplier records, quality results, service reports, end-product manufacturer, and final-customer complaints to detect, analyze, and eliminate causes of probable deficiencies (RBPS).

- **Trend Analysis.** Trend analysis is a simple monitoring and analytical tool to determine if SCRM processes are stable, capable, and improving. Trend analysis can detect pending or potential risks or problems. Typical examples of risk or problem trends include high rework, customer service reports, high product waivers, continued inspection, and high waste levels.

Trend and data analysis reveal audit, corrective action, and preventive action effectiveness. If supply risks, deficiencies or problems continue, this indicates the symptom was fixed but the root-cause of the problem was not discovered and eliminated. When a risk, deficiency, or nonconformance (defect) is discovered, the problem and data are entered into a database for future examination. Did the problem recur? How often and in what area did it recur? Could the recurrence be due to multiple factors (RBPS)?

SCRM IMPROVEMENT

The difficult we do immediately. The impossible takes a little longer.
Anonymous

The traditional purchasing model is the 'across the table', arms-length transactions. There are several problems with this model. Each party distrusts the other. Each party is transaction oriented. Each party tries to maximize the benefits they would receive from the deal. Each party also may work according to its set of preconceptions. The end-product manufacturer thinks suppliers will over-charge, over-promise, and under-deliver. The supplier thinks that each purchase is the last and will try to maximize the deal. So, buyers play suppliers against each other, switch suppliers on a dime, and offer one time or short-term contracts. The basic business model is short-term and adversarial. Supply development and innovation requires a new mindset (RBPS).

SUPPLIER INDUCEMENTS

SCRM is a new business model. In other words, end-product manufacturer and key supplier processes are complementary and are integrated into a seamless chain. To do this, suppliers are induced to change, adapt, and adopt new SCRM practices.

There are several methods by which to do this. One simple inducement is to ensure that supply-partners are paid in cash with a new purchase order. Lower level suppliers are paid net 30 days.

Being 'designed in' or 'sole-sourced' into present or future products can also induce suppliers. The supplier's products are then locked into a long-term contract and the supplier has a greater potential for future business. This type of 'sole sourcing' is used sparingly because the risks to the end-product manufacturer can be high.

Supplier rewards and inducements may include:

- Longer-term contracts.

- System contracts.

- Technical assistance.

- Public recognition of superior supplier performance.

- New opportunities for additional business.

- Supplier awards.

- Early involvement opportunities in product development.

KEYS TO LONG-TERM SCRM RELATIONSHIP

The transactional purchasing model evolved when SCRM stakeholders realized that there could be a better way to do business. Adversarial relationships could cause constraints, roadblocks, bottlenecks, and lost opportunities. If a supplier had the right attitude and energy, these could make up for other deficiencies.

An end-product manufacturer wants to work with a supplier that has the right attitude to improve and collaborate. This applies to first-tier through lower tiers. Supply development often starts with a first-tier supplier. Joint rules, plans, and forecasts are critical to customer-supply partnering. Real collaboration between companies involves more than sharing information. It is an integrated process that involves joint planning, deployment, and measuring results.

LONG TERM RELATIONSHIP
The following are critical for maintaining a long-term, customer-supply relationship:

- Open and honest communication.

- Trusting relationship.

- Mutually beneficial relationship.

- Continuous innovation.

Open and Honest Communication
The key to maintaining a great customer-supply relationship is open and honest communication that anticipates difficulties and establishes trust. The end-product manufacturer keeps the supplier informed of order changes and design modifications. The supplier keeps the end-product manufacturer informed of changes in delivery dates and production problems. Clear communication settles issues before they become problems or disputes.

Trusting Relationship
Trust is the hopeful outcome of clear communication. If the end-product manufacturer and supplier trust each other, this serves as the basis of a mutually beneficial relationship. The end-product manufacturer consistently receives defect-free products and knows the supplier will work toward improving product reliability, controlling risks, and lowering costs. And the supplier obtains a long-term contract.

For SCRM to work properly, both end-product manufacturer and supplier must demonstrate:

- Mutual benefits are real, tangible, and quantifiable.

- Partnership is based on a formal and binding contract.

- Each party meets its obligations.

- Continuous innovation of cost, quality, deliver and service is pursued.[157]

Mutually Beneficial Relationship

The end-product manufacturer tries to establish a mutually beneficial long-term relationship. It is easier to maintain an existing relationship than to start one from scratch. If a supplier has been providing a unique product at a reasonable price with good service, then it makes good business sense to keep and develop the supplier.

Partners need to share new-product development information. Developing goodwill and trust takes time and effort. Firms can generate trust by exchanging critical technical resources; sharing confidential technology forecasts and drawings; collaborating on strategic plans and long-term product forecasts; or exchanging sensitive financial information.

Continuous Innovation

Suppliers are asked to continuously improve product risk-controls, quality, delivery, technology, and service. In other words, they are asked or required to move up the capability and maturity curve. At a minimum, defect prevention hopes to eliminate routine incoming material inspection. If a supplier can demonstrate a history of defect-free shipments, the supplier is audited periodically to ensure internal process controls are followed and documented.

At this point, shipments go directly into production. Supplied product samples are periodically evaluated or certificates of compliance reviewed.   If successive shipments have been accepted and the supplier can prove internal process controls are in place, then a supplier becomes certified and all shipments are sent directly to the end-product manufacturer's production line. If a shipment is rejected, then shipments are immediately canceled and problems are root-cause resolved (RBPS).

Auditing and corrective action are probably the two most critical processes for improving supply performance. They 'close the loop' with all supply chain problem solving (RBPS). They form the basis of monitoring, intervening, correcting, and improving most customer-supply processes (RBPS).

MANAGING CUSTOMER-SUPPLY DIFFERENCES

Successful collaborative negotiation lies in finding out what the other side really wants and showing them a way to get it, while you get what you want.
Herb Cohen

End-product manufacturer and supplier differences will always arise. How these are corrected and prevented from recurring is addressed early in the customer-supply relationship. Little differences in opinion and expectations can cause irreconcilable problems between the end-product manufacturer and supplier. If problems arise, then immediate follow up, joint problem solving (RBPS) and root-cause solution can cement the customer-supply relationship. Even in the best of SCRM relationships, problems can arise. These challenges are opportunities to build long-term trust (RBPS).

PROBLEMS AND RISKS CAN ARISE ANYWHERE, ANY TIME

We often think the source of a problem is the supplier. But, either the end-product manufacturer or supplier can be the source of the problem. The end-product manufacturer can modify a drawing so the supplier can no longer make the part. The end-product manufacturer may want a sudden increase of products shipped to the plant, but the supplier is already running three shifts at capacity. The end-product manufacturer may tighten specifications and tolerances so the supplier cannot make the product. This occurs when the end-product manufacturer wants improved quality and reliability too quickly and the supplier does not have the internal capability to make products to specification (RBPS).

The supplier can also have problems. A supplier is on strike or can have fire, capacity, financial, or even supply problems and risks of its own. Any of these can disrupt or stop the supply chain (RBPS).

WHEN IT HITS THE FAN!

There are early indicators or precursors of supply chain trouble and risk. Costs are high. Supply chain is unstable. Processes are not capable. In-

---

CONTEXT: Warning Signs of a Risky Supplier

- Supplier requests for price increases, accelerated payment terms, customer financing support, or use of factoring.
- Late deliveries or changes in product quality.
- Requests for technical support.
- Failure to update IT systems or to appropriately use existing technology in the industry.
- Failure to effectuate cost reductions.
- Deteriorating accounts receivable and accounts payable.
- Employment of consultants and financial advisors.
- Deteriorating market position.
- Restatement or delays in issuing audited financial statements.
- Changes in key management positions.
- Renegotiated debt covenants, incurrence of new debt, fully drawn lines of credit and impending maturity dates.[158]

---

ventories creep up. Stakeholders are unhappy. Final-customer satisfaction is low. Field failures are high. It is critical to quickly understand the supply chain problem and pinpoint its root-cause. Then a supply management team is assembled.

It is assumed that partners are voluntarily participating and have agreed to jointly solve/fix problems. However, a sub supplier (second-tier or lower) in a sub chain may not be as disposed as a first-tier supplier to readily root-cause fix problems. We call this a risky supplier (RBDM).

A risky supplier is one who is unresponsive. A risky supplier is thus monitored or audited more closely than other suppliers. Close monitoring may involve more frequent plant visits to intensified process monitoring. Close monitoring usually ends when the supplier has shipped acceptable products over a period. If nonconforming products continue to be shipped, the supplier is trained, induced, or removed from the approved supplier's list.

CORRECTION AND PREVENTION

Problem solving and constructive innovation are what business is all about.
Randall Meyer

If the audit finds no deficiencies, then the audit findings report this. If the

---

CONTEXT: FDA Corrective and Preventive Action Regulations

Each manufacturer shall establish and maintain procedures for implementing corrective and preventive action. The procedures shall include requirements for:

1. Analyzing processes, work operations, concessions, quality audit reports, quality records, service records, complaints, returned product, and other sources of quality data to identify existing and potential causes of nonconforming product, or other quality problems. Appropriate statistical methodology shall be employed where necessary to detect recurring quality problems.
2. Investigating the cause of nonconformities relating to product, processes, and the quality system.
3. Identifying the action(s) needed to correct and prevent recurrence of nonconforming product and other quality problems.
4. Verifying or validating the corrective and preventive action to ensure that such action is effective and does not adversely affect the finished device.
5. Implementing and recording changes in methods and procedures needed to correct and prevent identified quality problems.
6. Ensuring that information related to quality problems or nonconforming product is disseminated to those directly responsible for assuring the quality of such product or the prevention of such problems.
7. Submitting relevant information on identified quality problems, as well as corrective and preventive actions, for management review (RBPS).[159]

---

audit finds deficiencies then corrective and/or prevention actions are recommended. Corrective and preventive actions are essential elements of all supply chain problem solving (RBPS). Corrective and preventive actions ensure symptoms as well as their root-causes are eliminated so supply problems do not recur.

WHAT IS CORRECTIVE ACTION?
Corrective action may involve:

328   Chapter 12: SCRM Risk-Assurance

- Architecting, designing, deploying, and assuring additional risk-controls.

- Redesigning or modifying the product.

- Redesigning or modifying fixtures and dies.

- Training suppliers.

- Updating specifications or changing tolerances.

- Increasing incoming material inspection.

- Containment of nonconforming products.

WHAT IS PREVENTIVE ACTION?

According to the ISO, corrective action is any "action to eliminate the cause of a detected nonconformity or other undesirable situation. Once the root-cause of the nonconformity is discovered, corrective action is undertaken to eliminate the recurrence of the risk, defect, flaw, problem, or discrepancy (RBPS)."[160]

Corrective and preventive action is sometimes confused. Corrective action is the process of eliminating a nonconformance and ensuring that it does not recur. To do so, the operations manager or process owner must understand the nature of the problem and suggest a solution such as sourcing a new fixture, developing a new procedure, or training employees.

Corrective action results from recurring end-product manufacturer and final-customer complaints, product nonconformances, unstable processes, or product rework. On the other hand, preventive action attempts to anticipate a potential supply chain problem and identifying steps to eliminate its possible occurrence. How can potential supply chain be anticipated? One method is to identify high risk areas and ensure there are internal controls to minimize or eliminate the risk (RBPS). Another method is to use the 4P's method discussed earlier.

CUSTOMER-SUPPLY COMPATIBILITY

'Culture trumps strategy' is a famous quote. We have been risk auditing suppliers for many years. Most supplier audits tend to focus on the supplier's ability to produce products within quality requirements or the financial stability of the end-product manufacturer. But, we have found that there is a more critical customer-supply factor: cultural compatibility.

Change initiatives fail unless there is a cultural fit. If there is a cultural interference fit, there is a high probability that the SCRM initiative will fail. Bottom-line: strategic sourcing or supply chain management will succeed if there are unity of purpose and shared wins/rewards among stakeholders.

Can both partners understand and comply with the needs, wants, and requirements of the other party? Many companies assume that if the supplier has good designs, stable manufacturing, reliable quality controls, and has been in business for years, then problems can be corrected and differences smoothed over. Often, this is not the case with new offshore suppliers. And in some cases, the culture gaps result in finding different or even domestic suppliers (RBPS).

SETTLING DIFFERENCES

Put a good person in a bad system and the bad system wins, no contest.
W. Edwards Deming

So, what can be done with problem or recalcitrant suppliers? Dispute resolution systems and procedures have to be established early on so differences can be readily settled.

OPTIONS AND RESOLUTION

End-product manufacturers can pursue the following options to resolve supplier conflicts:

- Work with existing suppliers.

- Manufacture component in house.

- Find new suppliers.

Work with Existing Suppliers

If more products, tighter tolerances, lower prices, and faster delivery are needed, the preferred option is to work with existing suppliers. Existing suppliers already know end-product manufacturer needs and wants. If the end-product manufacturer anticipates the need for more products and the supplier is running at full capacity or the supplier cannot manufacture products to specifications then the end-product manufacturer may provide the supplier with financial and technical incentives.

Manufacture Component In-House

A less preferred option is to develop the internal capability or competency to manufacture the component. This is done if the supplier cannot consistently make conforming products because of material, technology, cost, or personnel problems. Critical or state-of-the-art products may require additional investment in property, plant, equipment, and extensive personnel training (RBPS).

Find a New Supplier

The third option is to find a new supplier. The process of evaluating and selecting a new supplier was covered earlier. The process is long and tedious, so it is a less preferred than inducing, training, and negotiating with existing suppliers.

Finding a new supplier is difficult. Think of all the things that are done to establish the level of trust that existing suppliers have. That is why end-product manufacturers prefer to help their existing suppliers improve and mitigate risks rather than locate, qualify, and switch to alternates.

Customer-supply collaboration implies both parties will attack problems through joint problem-solving teams, will identify root-causes, will experiment to identify sources of variation, and will deploy supply chain innovations. Benefits are then shared equally between the end-product manufacturer and supplier (RBPS).[161]

As we discussed, end-product manufacturers will change suppliers when the pain of changing suppliers is less than working with a supplier. What would induce an end-product manufacturer to look for a new supplier? The end-product manufacturer may want to:

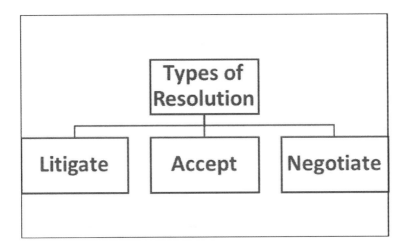

Figure 35: ***Types of Resolution***

- Lower customer-supply risks.

- Improve supplier delivery, cost, and quality.

- Reduce inspection or rework.

- Lower prices and overall costs.

- Improve supply metrics/objectives.

RESOLVING CUSTOMER-SUPPLY CONFLICTS

Many of our best opportunities were created out of necessity.
Sam Walton, founder of Walmart

Supply chain conflicts always arise and can be resolved in three ways:

- Litigate.

- Accept.

- Negotiate.

LITIGATE
The least-preferred option is to rely on lawyers and the courts to settle differences. Litigation is expensive and requires time to decide and, unfortunately, the outcome is usually ill will between parties.

ACCEPT

Another option is for the end-product manufacturer to accept the problem or risk. Not resolving disputes or problems in the short-term results in larger, long-term problems. A TV commercial sums up the situation by warning 'pay me now or pay me later.' When problems festered for years, the only way to resolve a dispute is through litigation and the cost of litigation can threaten a firm's survival (RBPS).

NEGOTIATE

The preferred option is to negotiate or arbitrate differences. Differences then are resolved quickly and inexpensively. A win-win partnership can slowly evolve where the end-product manufacturer gets quality products delivered just-in-time and the supplier gets a long-term contract.[162]

More end-product manufacturers are redesigning their internal and supply chain processes to deliver value adding products and services to their final-customers. First-to-market and time-to-market are competitive differentiators. Product development or time-to-market is too long for many companies.

There is usually a wakeup call to transform the product development process. Automakers recognized there was trouble when Japanese automakers were consistently designing and building a new automobile in less than 30 months when the U.S. 'Big Three' automotive manufacturers required 48 to 60 months to accomplish the same set of tasks.[163]

FOOD SUPPLY RISK-ASSURANCE

In the United States, when 1 in 6 Americans or 48 million people get sick, 128,000 are hospitalized and 3000 die due to food borne illnesses, there is going to be more scrutiny of farm food chains.[164]

Over the last 5 or so years, food, pharmaceuticals, electric power, electronics, and many sectors have developed their own SCRM risk-assurance requirements. One of the more visible sectors moving to risk management involves food. Active management of the cost, delivery, quality, and risk of supplied food is now mandated and statutory in many cases.

CLIMATE CHANGE

What is the driver of food SCRM? Climate change and food disruption already create political instability according to many reports. One report recently warned of a "growing risk… the human security," "systemic disruption," "Black Swan events."[165] Climate change or global warming is expected to produce more frequent and more violent storms resulting in increased flooding in many areas. Temperatures will rise, rainfall will change, and weather volatility will increase, which will impact crop yields.

The report entitled 'Choke Points and Vulnerabilities in Global Food Trade' describes how climate change will impact national infrastructure:

> "In addition to more regular and more severe weather-induced damage to roads, railways, ports and inland waterways, climate change will have a multiplying effect on security and political hazards affecting the infrastructural backbone of international trade."[166]

These choke points have the potential of being "potential epicenters of systemic disruption,' which will impact all supply chains.[167] The solution is that populations must become more resilient as cascading supply chain disruptions become more common.

CLIMATE CHANGE AND NATIONAL SECURITY

Climate change will also result in mass migrations, government failures, wars, ungoverned areas, and supply chain disruptions. We are already getting a glimpse of these events with the mass migration of refugees to Europe, ungoverned states in the Middle East, and more localized conflicts, and random acts of terror.

Climate change is also impacting national security. A recent report distilled the strategic risk of climate change:

> "The effects of climate change present a strategically-significant risk to U.S. national security and international security."[168] … "U.S. must advance a comprehensive policy for addressing this risk."[169]

FOOD SUPPLY CHAIN RISK-ASSURANCE

Climate change will also impact food supply chains. For example, about 2.25 billion cups of coffee are consumed daily. Coffee is the second most valuable commodity exported by the developing world worth approximately $19 billion. A recent study concluded that climate change may cut coffee production by as much as 50% over the next 20 years resulting in supply shortages and increased prices.[170]

The growth and delivery of food is a global, complex system especially with rising global populations. Unfortunately, complexity can result in risks such as food contamination and food borne illness factors. Food contamination may include listeria, E. coli, and botulism.

The food supply chain is a global supply chain network that is coming under more scrutiny with higher risk-assurance requirements. The challenge is food outbreaks can result in food recalls with increased media coverage and heightened consumer awareness resulting in higher government regulation and oversight. Food contamination already has had a material impact to organizations such as McDonald's, Burger King, Subway, Taco Bell, and Chipotle.

Food contamination has resulted in multiple class-action lawsuits. These lawsuits can have massive impact on the company's reputation such as for Chipotle or can be a major nuisance claim such as objections for genetically modified food products.

FOOD SAFETY REGULATION

In the U.S., the Food and Drug Administration (FDA) regulates the production and safety of food. As we discussed, there is now more concern about food safety largely because of incidences such as e coli in salad, listeria tainted cheese, salmonella contaminated peanut butter, meat adulteration, and even toxic ingredients in common foods. Some of this is due to carelessness in the shipment of food, but more often food handling is the greatest risk challenge due to food deterioration.

Let us look at the food supply chain, which is critical for public safety. In response, the U.S. Food and Drug Administration (FDA) has developed standards for the manufacturing, sorting, and testing of food. The FDA is concerned about contaminated or adulterated ingredients and food that can have wide impact on public safety. Food manufacturers and service organizations are concerned because of the exposure they have in terms of liability and potentially losing brand equity.

Now, food companies must deal with a wide range of regulations. Food companies are aware they need to develop more supply chain risk-controls especially as the length, complexity, and uncertainty of the food supply chains impact thousands of products and hundreds of suppliers scattered throughout the world. The results are frightening because of multiple areas of risk, threats, and vulnerability. The solution is to develop SCRM through-out the different food supply chains.

SUPPLY CHAIN RISK MANAGEMENT

Food products are categorized based upon levels of risk. For example in the food sector, seafood, fresh produce, and meat are high risk products. Medium risk food products include raw produce. Low risk products include foods that have not been effectively processed but do not support pathogens.[171]

The fish supply chain also has high exposures of risk. It is one of the most critical supply chains since tainted fish can cause widespread health concerns. So over the last few years, each type of fish related food has been screened for safety. Why? One in five fish tested worldwide is mislabeled. Food risks can increase in terms of eating contaminated, mislabeled, illegally caught, or fraudulently substituted fish. Each of these can result in consumer and organizational risks.

The U.S. government is now requiring traceability of 13 types of fish to deter fraud and mislabeling that occurs within the seafood chain. Some believe that traceability may lead to information about the fish species, how the fish are caught, and even auditing the supply chain trail.[172]

FOOD SAFETY MANAGEMENT ACT

The U.S. Food Safety Management Act (FSMA) was signed into law in 2011 and focuses on preventing food safety issues including controlling supply chain risks. Under FSMA, food importers must identify foreign food suppliers and determine if they are compliant with FDA statutes and regulations. They also have to conduct a hazard analysis on all imported food products to understand the risks associated with suppliers and ensure those risks are managed effectively. An important element is auditing supplier facilities, testing its programs, and reviewing supplier food safety programs.

Under FSMA, the FDA has proposed the Food Supply Verification Program (FSVP). The statute puts the responsibility of controlling supply chain risks onto the brand owner or the end-product manufacturer of the food.

Food Risk Levels
Food importers must now determine if the following risk levels associated with imported food may exist:

- Review inherent food safety hazards associated with the food, such as chemical, biological, recall history, etc.

- Review tiers of the imported food supply chain and report number of suppliers, complexity of the supply chain, supplier history, supplier compliance, manufacturing details.

- Assess risks associated with production, packaging, transporting, and storing the food.

- Review supply chain macro factors, such as country of origin politics, natural disasters, local food culture and other factors

Food Assurance
Regulations also require food verification and risk audits of imported food suppliers, which may entail:

- **Onsite audits.** When there is a reasonable probability that exposure of the hazard may result in serious negative health consequences.

- **Inspections.** Inspectors may review food facilities, processed, fish, and controls.

- **Sampling and testing of food.** Periodic sampling and inspection of food provides risk-assurance of food safety. Inspection is increased or decreased based on the required risk-assurance.

- **Review of the supplier's food safety records.** Food manufacturer are required to maintain records and evidence of inspections and food controls.

- **Certificates of Analysis (COA).** The end-product manufacturer may require a Certificate of Analysis (COA) to demonstrate evidence that suppliers have adequately tested the product or ingredient against a specification. The idea behind a COA is an independent third-party will conduct tests in a certified laboratory to provide at least 95% assurance that products comply with FSMA and other specifications. This is becoming more common in many sectors. This is a form of risk sharing where the laboratory assumes part of the risk by conducting an independent test in an approved and competent laboratory.

SUMMARY
Supply management several years ago was a safe function within an organization. Demand was stable. Parts were pulled through the chain. Supply-partners were well known. Now, things have changed radically. In 'What Does It Take to Shut Down a Supply Chain?' the author says: "All it takes is just one glitch to bring a supply chain to its knees"?[173] The one glitch could be a bad decision (RBDM).

The relationship between the end-product manufacturer and its suppliers preferably is a partnership and collaboration. The reality is there is a fine balance of power and profitability. To make a true partnership or collaboration work, the end-product manufacturer must work with select suppliers to assure everyone's requirements are understood and are met.

An audit, particularly a quality management system audit, is a snapshot of a supplier at a given point in time. The challenge is the supply chain is really a moving picture. The auditee (person or area being assessed) may present an over optimistic image that does not reflect supply chain realities. On the other hand, a minor risk or problem is blown out of proportion and may not accurately reflect the general conditions of the supply chain (RBPS).

NEXT CHAPTER

In the next chapter, we discuss SCRM innovation.

CHAPTER 13:
SCRM INNOVATION

What makes a successful company? Innovative products and stable supply chain processes allow a company to bring products to market quickly and generate revenue. Steve McConnell, the author of **Rapid Development**, says there are "10 to 1 differences in productivity between companies within the same industries" that develop state-of-the-art processes like SCRM.[174]

THE COMPLEXITY OF BUSINESS
As business and life become more complex, end-product manufacturers are losing their ability to compete by following their present business models and practices. This delta, the growing distance between 'current practices' and 'SCRM innovation and best practices' is the zone where operations managers can innovate and excel. This zone can be seen in the figure on the next page and is where there is the greatest potential to implement SCRM and enhance value.

INNOVATIVE COMPANIES WIN
Fast Company, the innovative business magazine, says to win in business a company has to work smarter not harder. The logic goes like this. If a company wants to win at business, it needs the best people and the best business models and processes, such as SCRM.[175]

The *Economist Magazine* also made the case for innovative work processes:

> "Innovating has become the buzzword of American management. Firms have found that most of the things that can be outsourced or reengineered have been (worryingly, by their competitors as well).

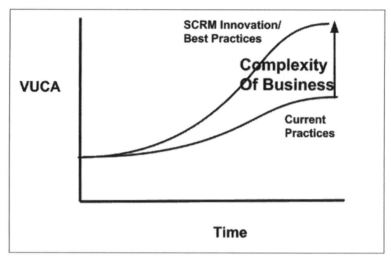

*Figure 36: **Growing Complexity of Business Due to VUCA***

The stars of American business tend today to be innovators, such as Dell, Amazon, and Walmart, which have produced ideas or products that have changed their industries."[176]

The financial numbers reinforce the power of innovation. The top 20% of firms in a *Fortune Magazine* innovation poll had double the shareholder returns of their peers. And, that frightens all end-product manufacturers. It is all about killer ideas these days – ideas that can disrupt processes, projects, people, organization, organizations, sectors, and ultimately society.[177]

PRODUCT DEVELOPMENT MATH

Some end-product manufacturers attempt to develop a new generation of products yearly. The challenge: some 13,000 new products hit the market each year, but only 40% will be around 5 years later.[178] The time-to-market math is pretty Darwinian.

Examples of rapid product innovation can be found all over. Hewlett-Packard wants 80% of its revenues to come from products that are less than 3 years old. Fashion designs change quarterly or even weekly. Software is enhanced yearly. Twice as fast computer chips are introduced every eighteen months. Even in the automotive industry, General Motors wants to halve product development time from the current minimum 40 months to 24 months or less.[179]

MASS CUSTOMIZE OR NOT

Another conventional wisdom in product development seems to have shifted. Is it true the final-customer wants freedom of choice in all things? We are offered many options and features in software and gadgets. The reality is that we are confused and do not use the additional features. Several examples illustrate this. At home, I have a control device for my VCR, TV, stereo, etc. Each device has 10 or 20 control features. None will talk to each other. I use the on/off, volume and selection changes. That is it! The same is found with software. I use the most popular software office suite and use probably 2% to 5% of its features. What is the benefit of the rest of the features?

Well techno-wizards are getting smarter and have discovered that we do not want the extra bells and whistles or unlimited choice. We want basic value. Not the 'one size fits all' product or service but a good product that is designed to satisfy our most needed requirements or in other words a simple mass-customized product. For example, MCI or phone carrier may call you, review your account, discuss your options, look for potential savings, and present one or two choices. Again, the purpose is to offer value and make the buy decision simple (RBDM).[180]

DESIGN CHALLENGE: LISTEN TO THE FINAL-CUSTOMER

The history of technology is full of neat ideas that did not work out. Sometimes, the final-customer had unrealistic expectations. Sometimes, the development team was clueless of what the final-customer wanted. Sometimes, they did not connect. These techno-turkeys included all types of products and inventions.[181]

The challenge is to make technology seamless with work or in other words to make technology disappear for the user. People do not need to decipher manuals or write code. The development team needs to think how the supply chain is integrated to design 'cool', customer-friendly products.

SCRM INNOVATION

FISH: First in. Still Here.
Anonymous

The goal of supply chain innovation is to provide final-customers with the 'right' product at a reduced cost, while having it easily accessible. How is this done? Mass customization is one method, which says that a product will be designed and tailored to the needs of the individual customer. In other words, products are made to specific final-customer requirements with defined value attributes.

WHAT IS MASS CUSTOMIZATION?

Mass customization is a marketing philosophy that is a major SCRM driver. Mass customization implies that by using a standard product, different customers can be provided with specialized products. Specific customer value-added features such as product feel, functionality, and bells/whistles are added to the basic product to satisfy different customer segments and requirements.

The challenge is that mass customization requires supply chain flexibility. Manufacturers may have to build 20 or more different products using the same production line one week and 10 or more different products on the same line the following week or next day.

THE $7 TOOTHBRUSH

In the hyperactive toothbrush market, manufacturers are mass customizing by adding features for different final-customers based on a common brush platform. They design bristles of varying lengths, add new materials, and provide flexible handles. And instead of focusing on cavity prevention, the new brushes allegedly lower gum disease.

Would you pay $7 or more for a new premium toothbrush, which is a disposable, commodity product? Manufacturers from razors, toothbrushes, and other disposables bet you would. It is the most expensive mass-market toothbrush ever. What is different about it? It offers several critical innovative, final-customer value attributes. It has three types of multi-colored bristles set at different angles. The dense tip cleans behind back

---

CONTEXT: Dell Build to Order (BTO)

Dell is often cited as the poster company with the best BTO processes. Their BTO business model works this way:

A customer orders a computer directly by phone or by email. The customer only has a few computer options and features. The customer pays upfront by credit card. The computer is then assembled and sent within 3 days of the order.

The secret to Dell's BTO is to design standardized systems and modules that are configured and assembled quickly. Dell suppliers provide preassembled systems and modules on-demand. Dell restricts customer choice to a few modules and options that are assembled within 4 minutes of the customer order. Installing the software takes about 90 minutes longer. Most of the customer customization comes from the software that is chosen by the customer.

By leaving customization to the end of production, Dell is able to mass-produce computers while leaving the 'customer of one' customization to the end of the buy process.

---

teeth. The handle is rubberized and ergonomically designed. It looks and feels cool. And, it sells for 50% more than its traditional high-end rivals.[182]

What is happening? Motivating customers to trade up to a new generation of products through mass-customization in a seemingly mundane product is a highly profitable strategy. A company can get both high volume and high profit margins simultaneously. Companies may develop a new product or jazz up a consumer product and charge a premium for them. Nike did it with its $200 sneakers. Starbucks did it with its $4 to $5 exotic coffee blends. Gillette did it with its Mach 3 razor.

SUPPLY CHAIN CHALLENGES
To mass customize, four supply chain challenges have to be addressed and solved (RBPS):

- **Smaller lot sizes.** Reducing process 'change over' and machine 'set up exchange' times become critical to producing different products in different lot quantities.

- **Information Technology.** More devices are smart with an Internet address and accessible via online. As well, these smart products offer more functionality with features that increase exponentially. As smart products increase, the variation of product attributes, costs, and delivery information increases to the point where sophisticated computerized networks are required to monitor throughput, plan/forecast demand, store information, and anticipate supply chain bottlenecks.

- **Short product cycles.** Short cycles make it difficult to fine tune production processes, much less supply chain processes. Supply chain process and product spikes create supply chain bottlenecks, constraints, exceptions, and nonconformances (product).

- **Complex SCRM business relationships.** Stakeholder and interested party relationships in the supply chain become more complex. Customer-supply relationships were once transactional and product based. Now, they are process and risk based, incorporating process competencies of the supply chain partners. More advanced SCRM relationship are based on strategic ERM partnerships.[183]

PRODUCT INNOVATION

In the future, more people will work for themselves, creating a huge market for bizarre products.
Scott Adams, 'Dilbert' creator

Operations managers and supply-partners should be introduced early into product development.

CONCURRENT PRODUCT DEVELOPMENT

To shrink development cycle times, critical product development activities are concurrent instead of sequential. In concurrent product development, design participants including supply-partners can interact in real-time, sharing designs, bills of material, and other design images on a computer monitor. The development team uses computers to simulate user requirements, design in 3D, do 'what-if" analyses, and test products under different conditions. Everyone sees the same documents in real-time. Documents and drawings can be discussed and amended during the meeting. Iterations, redundancies, and costs can be eliminated as work is simplified.[184]

While linear product development was time consuming, it offered benefits. The process worked because it was predictable, redundancies ensured that mistakes were caught, and all parties were likely to understand requirements.

Concurrent product development is messy. Communications are more difficult. There are more opportunities for mistakes late in the product life cycle and it is difficult to break the 'toss it over the wall' development mentality.

EARLY SUPPLY INVOLVEMENT

Critical supply chain stakeholders including supply-partners are brought in early in product development. Supply chain involvement in the design process may be as simple as resolving design conflicts such as aligning holes in a product or ensuring parts are accessible for maintenance.

More often, operations managers are familiar with design technologies, specifically computers and other electronic tools that make cost-effective design possible. Engineers use computers to understand and change designs much like writers use software to move words and paragraphs to develop a book.

STORIES OF INNOVATION

I think frugality drives innovation, just like other constraints do. One of the only ways to get out of a tight box is to invent your way out.
Jeff Bezos, founder of Amazon and a Supply Chain Innovator

The following stories of innovation illustrate the importance of supply-partnering:

BOEING 777

Product development now utilizes the best people, principles, and practices in virtual teams. Design data is transmitted real-time to stakeholders so the virtual team can collaborate on designs. Boeing's engineers and supply-partners designed the 777 jet using a computer-aided-system to develop and assemble a virtual plane so hundreds of thousands of parts fitted when the first prototype was assembled.

TOYOTA

Toyota says it can produce a car within 5 days of a custom order. This is startling because the auto industry has often kept end-product manufacturers waiting 30 to 60 days. Why is this important? This is a huge competitive advantage in today's Internet economy, as end-product manufacturers want instant online order verification. Faced with a long delay, most U.S. end-product manufacturers will settle for what a dealer has on the lot.

How does this work? Toyota auto planners have developed a 'virtual production line.' The system calculates exactly which supplied parts need to be available at each point of the production line so the expected mix of vehicles is determined days prior to actual production.[185]

SWATCHMOBILE

Have you seen Swatches? They are funky, cheap, colorful, and changeable watches. A person can change watches three or more times a day depending on his or her moods, functions to attend, or attire. Swatches for some have evolved in a living and working accessory.

People can now update their automobiles similarly. Daimler-Benz engineers recently designed a go-cart sized, plastic vehicle called the Swatchmobile to satisfy fickle consumer wants, test-design plastic bodies, and develop supply-partnerships. Final-customers can change the 'look and feel' of their vehicles on a whim and at a nominal cost. The ultra-light, ultra-fuel efficient two-seater sells for about $8,500.

Swatchmobile suppliers do not just make handles and headlights. They supply the entire door, front end, or cockpit as modules to be assembled. The suppliers also install the parts so a Daimler operations manager is little more than a product coordinator where most employees are even on someone else's payroll.

MATURATION OF SMART PRODUCTS

Now, we have smart phones, smart cars, smart medical devices, and other smart things. And, smart things are moving quickly from industrial products to retail and now to consumer products.[186]

The basis for much of the Internet of Things is the interconnection of products using the Internet. These products have an IP address. Using a smart phone, a user can visually monitor and control products.

Already, there are smart products that can diagnose vehicle conditions, physical condition of humans, and vehicle failure monitoring in real-time, such as when the car is moving. Traditionally, automotive analysis was only available at the dealership when the computer was connected to the vehicle.

MEDICAL DEVICE RISK

Take a look at medical device security. This is a life or death concern. End-product manufacturers simply do not have the expertise to design and mitigate all the possible IoT product vulnerabilities in infusion pumps, heart monitors and internal medical devices.

In a recent survey, almost 1/3 of the professionals indicated that "identifying and mitigating the risks of fielded and legacy connected devices is one of the medical device industry's biggest cybersecurity challenges."[187] The second concern was the possible "embedding vulnerability management into the design phase of the medical device." [188]

NISSAN

Nissan recently launched a vehicle monitoring system that can predict automotive component failure and automatically notify the owner and dealership that an automotive part is about to fail and should be serviced quickly. Now, that power is given to the owner. As part of the smart vehicle with smart components, the connected vehicle will alert the driver via smartphone. The monitoring system can also identify defects in engine

---

CONTEXT: Ten Design Practices to Add Business Value

1. From department silos to cross-functional teams.
2. From narrow experts to interdisciplinary designers.
3. From cubicles to garages.
4. From a design stage to continuous design.
5. From qualitative to full-spectrum research.
6. From prototype once to prototype often.
7. From middle management to the C-suite.
8. From perspectives to metrics/objectives.
9. From financial to customer-based incentives.
10. From incremental to brave.[190]

---

components by assessing performance during acceleration and will alert the driver of the need for replacing a part.

INTERNET OF THINGS

'Is the Internet of Things Becoming an Internet of Risk?' was the title of a recent article. The premise of the article is 100,000 or more Internet connected devices such as baby monitors, refrigerators, door openers, thermostats, and other smart devices could be used in a distributed denial of service attack. This already has been done with Amazon, New York Times, and even Twitter.[189] Botnets attacks capture and control IoT commercial and even residential devices resulting in risks to servers, workstations, mobile devices, and even cloud deployment.

SMART APPLIANCES

Talking toasters, smart copiers, and thinking vending machines are coming to the supply chain. Toasters with a chip will recognize simple voice commands like light, dark, or burnt toast. When a smart copier or vending machine fails, or is about to fail and cannot self-correct, it will transmit an email to a technician to come and fix it. Service personnel will fix it just-in-time as it is about to fail. However, it will be a stretch for many of us to have a meaningful conversation with our talking toasters or washing machines.[191]

SUMMARY

Final-customer satisfaction will drive all business decisions. The *Economist Magazine* recently concluded the following:

> "Already, many companies find it more of a struggle than they did to win new customers and to keep those they already have. No surprise there: competition has sprung up from all sorts of new directions in the past few frenetic years and it will intensify as the downturn makes customers both pickier and more cautious."[192]

NEXT CHAPTER

In the next chapter, we discuss SCRM tools.

CHAPTER 14:
SCRM TOOLS

WHAT IS THE KEY IDEA IN THIS CHAPTER?

In this chapter, we discuss SCRM tools and techniques. Supply chains now are so lean and tight that any unexpected variability can totally disrupt the shipment of products and services. Supply chain risks can then manifest in unexpected capacity, delivery, quality, and cost variances.

SCRM TOOLS

SCRM has become an umbrella term for diverse operation excellence methods, principles, and ideas. Cycle time management, just-in-time (JIT) management, lean management, risk management, artificial intelligence, quality management, benchmarking, logistics management, and IoT. IoT especially is changing the nature of SCRM tools as the below indicates:

> "Certain steel and paper companies are building the Internet of Things directly into their production processes to detect or predict deteriorating quality. Their goal is to detect issues early enough for operators to 'save' the product by making adjustments in real time. Companies can set up product-specific rules to identify a quality defect (for example, the width or composition of tubes, paper, or steel plates) – and the faster they can recognize a quality problem, the earlier they can adjust manufacturing parameters or trigger maintenance tasks to head off a problem. This improves manufacturing output and speed to market, reduces waste and scrap, and minimizes energy for rework."[193]

Figure 37: **SCRM Best Practices**

CYCLE TIME REDUCTION MANAGEMENT TOOLS

Manage the opportunities change offers.
Advertisement

Cycle time reduction involves managing product development, schedule, and delivery risk factors. Time is a critical element of supply chain competitiveness. Since end-product manufacturer needs can change so quickly, the speed by which a supplier responds can affect the end-product manufacturer's reputation, credibility, and profitability. Cycle time management is the analysis, control, and reduction of how long it takes to do something critical. Thus, we are seeing the rise of fast fashion, personalized jeans, and other on-demand trends in almost every retail sector.

THE ACCELERATION OF CHANGE

Business hurdles and barriers are higher than five years ago. In only several years, low-price and high-quality products and services cannot guarantee success anymore. Products are developed and delivered quickly to satisfy fickle final-customer tastes. As well, risks throughout the supply chain are higher. End-product manufacturers want the right cost-competitive, customized products quickly.

Change is occurring at an increasing rate. What is the role of SCRM amidst this change? End-product manufacturers are struggling to streamline their internal supply management processes and upgrade the supply management function to reflect its new strategic importance.

THE POWER OF UPSIDE (OPPORTUNITY) RISK MANAGEMENT

Markets are fragmented with final-customers with diverse needs who want them fulfilled instantly with products or processes that not only exceed their expectations but may even astonish them. So, the first company often has a dramatic market advantage. The list of firsts is extensive – first-to-market, first to deploy SCRM processes, first to win the Baldrige Performance Excellence Award, first to implement just-in-time systems, and first to deliver a product to market.

An end-product manufacturer that can satisfy diverse final-customers with lower cost, high-value products and services delivered just-in-time and in the right manner has a higher probability of beating its competition. Delivering products and services faster requires the end-product manufacturer does the right things right the first time and every time. The overall value-added culture is ingrained in the end-product manufacturer. SCRM processes are standardized, simplified, proceduralized, stabilized, and capable. Everyone is trained to do their jobs right. Supply chain, non-value-added processes and systems are eliminated.

CYCLE TIME REDUCTION

Innovation and technology drive the supply chain model. Think Moore's Law, which says computer chip processing power doubles every 18 to 24 months. Now apply this to the supply chain. We are seeing this in almost every industry. For example, high-powered computers allow automakers to design two vehicles in the time they spent on designing one and do it in half the time.[194]

Motorola recently reorganized its semiconductor business because of increasing final-customer requirements. Traditionally, Motorola designed and manufactured generic computer chips for different markets. Now,

Motorola zeroes in on critical final-customer requirements so it can turn out a product in 30 days instead of three or more months.[195]

We have discussed fast fashion. Final-customers want on-demand fashion and do not want to wait 6 months to get the latest fashion shown in a magazine. There are multiple risks to retail brand owners. The offshore manufacturer may demand minimum order quantities of a fashion product that no one will purchase six months later. Then the retailer may have unsold inventory and working capital on the shelf not turning over. Ziel founder and CEO Marleen Vogelaar, an on-demand designer and manufacturer, said:

> "The average brand and retailer in the USA has 40% oversupply and 10% undersupply" … "That leads to about 20% write-offs and 30% to 40% discounts in the industry."[196]

So, many retailers are moving to made-to-order products, where guaranteed margins are lower but turnover is much higher.

CRITICAL CYCLE TIME REDUCTION AREAS

Cycle time reduction throughout the supply chain is critical to resolving chronic problems; improving customer deliveries; responding to fashion trends and tastes; eliminating waste; responding to product recalls; deploying new sourcing processes, and introducing new products to market (RBPS).

Time as an upside (opportunity) risk or as a competitive advantage is illustrated in the following examples:

Product Development Partnerships

Time-to-market, first-to-market, and first-to-critical-mass are key product development metrics/objectives. Few end-product manufacturers have the ability to create by themselves the products they need to be globally competitive and to satisfy all their final-customers.

Just-in-Time Production

Just-in-time (JIT) production or manufacturing is the systematic elimination of waste throughout the product or service delivery stream. Just-in-time starts internally and is then driven through the supply chain. The goal of JIT is to eliminate waste, lower operational risks, and enhance value. Superfluous raw material, extra material handling, inventory, extra labor, and inadequate supervision add unnecessary costs or risks to the process.

Just- in-Time Delivery

The marketplace changes quickly. New competitors arise. Final-customers are fickle. Final-customers want products delivered over night. In general, everything in business seems to accelerate at electronic speed. The windows of opportunity are shrinking as supply chain stakeholders are linked electronically. Parts are delivered just-in-time for assembly or test. Inventories are held down to close to zero. The time between an end-product manufacturer order and product delivery is reduced to a minimum. Products across the world are delivered within 24 hours of an order.

Large end-product manufacturers are now scanning the horizon for small companies with new ideas, lots of energy, and special skills they want but do not have. Sometimes, large companies will buy smaller, entrepreneurial companies. Sometimes, large companies will license technology or partner with smaller companies. These supply chain partnerships are called strategic alliances, joint ventures, strategic partnerships, or even virtual corporations. They exist as long as each partner adds value and as long as revenues are produced.

'PULL' OR 'PUSH' MODELS

Do you push or pull a chain?
Anonymous

Supply management metaphors abound. Is it a chain? Does it have DNA? Is it a well-oiled machine? Is it lean? Is it a 'push' or 'pull' model? In this section, we look at the popular just-in-time and push models.

SUPPLY CHAIN DISRUPTION

As end-product manufacturers move up the supply chain maturity and capability curve, operations managers want seamless and integrated processes from supply-partners. So when there is end-product manufacturer order, the designed product can be made to specific requirements.

Outsourcing, Six Sigma, lean, operational excellence, product customization, and other operational initiatives work well in a stable and capable environment. When there is supply chain disruption and volatility, these operational initiatives tend to break down. Why? These initiatives are often based upon assumptions such as lean where there are few or no buffer inventories. So, when there is a supply chain disruption and resulting volatility in shipments, there is no material to sustain production. The supply chain then grinds to a halt. Six sigma works when there are stable and capable processes. Disruption and VUCA are also changing the underlying assumptions of other operational excellence tools.

PUSH PROCESSES

For a supply chain to work smoothly, there needs to be a realistic estimate of product demand. Products flow through supply processes. As we have discussed, there are two types of flows: push and pull. And, there are various combinations between these two.

Traditional batch manufacturing follows a push model. Push demand starts from a sales estimate of how many products will be purchased by final-customers, which determines how many products will be produced. If the push forecast is too high, then extra products go into inventory. If the push forecast is too low, then extra shifts are used to produce the required products. Whatever happens, if the projections are too high or too low, there is a possible ripple or 'bullwhip' effect down and up the supply chain. Specifically, bottlenecks or high inventory risks can arise.

Anticipating and forecasting demand requires SCRM maturity that companies want to develop. This is an inherent risk to complex supply chains and fast design, where demand may not be forecastable. However, as the below quotation notes, if a company can develop accurate forecasting models, then this is a sustainable competitive SCRM advantage:

> "The companies best prepared to rapidly and effectively react to changing market conditions utilize modeling technology to create living models of their end-to-end supply chains, with the ability to redesign and re-optimize when forecasted changes or unplanned events occur."[197]

PULL (JUST-IN-TIME) PROCESSES

Pull demand systems are just-in-time and 'build to order' (BTO) processes. Final-customers create the demand, which may be retail or user consumption. This information flows up the supply chain to create the demand for additional products to flow down the chain. In a pull system, the final-customer defines what products are required and this determines product demand.

Build-to-order business (BTO) is a popular pull model and is used by many end-product manufacturers to reduce inventories. Dell early on saw the opportunities and patented many elements of its BTO business model. Amazon.com is also famous for its patented 'one-click' model – one mouse click ensures purchase of the on-line order. The unqualified success of companies such as Dell Computer, Amazon.com, and Cisco Systems, which carry little inventory, use their SCRM expertise as a competitive differentiator, and has driven many manufacturers to BTO pull systems.[198]

What is a better process, push or pull? This question has huge impacts and helps frame the SCRM discussion. Let us go back to the supply chain metaphor. It is powerful and simple. Do you push or pull a chain? We are hearing more about 'demand or pull chain' processes. The visual has some appealing qualities. It tells the world how the supply chain works. It implies every part of the chain is customer sensitive. And, the SCRM function has a key element to manage the process.[199]

RETAIL PULL SYSTEM

Technology has greatly increased the accuracy of estimating and predicting final-customer demand. Let us look at how Information Technology helps estimate demand in retail businesses. The traditional method of determining retail demand was the open shelf ordering process. The order clerk checked daily or weekly how many products had been sold and filled open shelves with inventory from the back room. If the order clerk over-projected demand in one store, then there was high inventory. If done in hundreds of stores, this could be catastrophic in the high volume, low-margin retail business. If the order clerk underestimated demand, there could be lost sales, unhappy bosses, and unhappy customers.

One solution is to pull products through production based on real final-customer orders. In this way, suppliers can produce products in small batches. Products move quickly through the supplier's plant. There is daily movement of products from the store to the consumer, from the manufacturer to the store, from the supplier to manufacturers and so on up the supply stream.

In many retail supply chains, product sales are electronically monitored and communicated to the respective suppliers. Consumer purchases generate demand information so products flow to where they are needed. This 'build-to-replenishment' business model much like the 'build-to-order' works as long as information is transmitted real-time to the distribution center, end-product manufacturer, and suppliers. Immediate order forecasts are developed with high accuracy and reliability. These serve as the basis for longer-range forecasts for product delivery and new product introductions months into the future. Again, this works well if there are stable market and demand conditions.

FROM DESIGN TO RETAIL IN 25 DAYS
E-commerce and online shopping are also based on a 'just-in-time' model. Online shopping has become a global phenomenon over the last 10 years. E-commerce has become such a big part of any business that in many areas it is more prevalent than brick and mortar retail shopping. In Canada for example, more than 76% of Canadians shop online anywhere from 4 to 10 times per year in 2015.[200]

Just-in-time manufacturing and fast fashion can work together to get the right product to the right market segments and final-customers quickly. End-product manufacturers working with key suppliers can react quickly to final-customer demands, changing requirements, and changing fashion trends. Lead times may change from quarterly fashion, to monthly and now weekly fashion.

ZARA JIT STORY
Zara is the world's biggest fashion retailer by sales and is the biggest 'just-in-time' innovator in the retail segment. Why? A story of a simple black dress with a metal clasp explains why.

In the world of retail, Zara has been in the forefront of fast fashion. Fashion used to be a quarterly design rush, where there were fall, winter, spring, and summer fashions. Now, fast fashion implies fashion cycles that are weekly and sometimes called 'see now – buy now.'

These are based on the just-in-time Toyota Production System that works equally well in fashion and retailing. Zara's fast fashion and "supply chain flexibility is what makes it four times more profitable than the average retailer."[201]

The rise of digital and social media has magnified the importance of SCRM. Consumer expectations and trends are shared almost instantaneously via Twitter and Instagram, so the fashion consumer who wants the black dress with the metal clasp can get it quickly. This is the future of retail and online shopping using faster customer gratification to generate sales.

BOEING MODEL
Boeing aircraft makes the fuselage of its planes and puts them in storage in order to meet spikes in demand or special airline requirements. Also, electronic suppliers have finished and assembled packages with components that are put in storage so they are customized for different end-product manufacturers.

CHALLENGES OF PULL AND PUSH MODELS
Bad things can and do happen to great supply chains. There can be poor replenishment of hot selling products. 'Out of stock' stickers make unhappy customers. Excessive inventories are expensive for everyone in the supply chain.

In a pure just-in-time system, there is no inventory. But, the reality is that the cost of a process hiccup, whether it is a partial delivery, lost shipment, or production slowdown can infect the entire supply chain and result in dissatisfied final-customers. All of which are costly. So, the operations manager and stakeholders must conduct a risk analysis and determine the probability of such an occurrence. If a partial or no delivery is unacceptable, then buffer inventory has to be used to manage and smooth out the production flow and supply stream. But buffer inventories are costly and anathema to lean management.

'BULLWHIP' RISKS

In a multiproduct, manufacturing environment, uneven demand and no buffer inventories can create additional risk just to keep a supply chain running smoothly and to keep the supply links coordinated.

The bullwhip effect is a common risk where final-customer demand forecasts are way off resulting in increasing inventory swings as one moves up the supply chain. The bullwhip effect refers to the increased amplitude of a bullwhip as it is cracked.

In the bullwhip effect, even small changes in a schedule at the end-product manufacturer's plant can lead to disruptions up and down the supply chain. If the entire supply chain process is not managed carefully, there can a bullwhip effect where production lots increase and transportation logistics are disrupted.

How can the supply chain risk of the bullwhip risk be mitigated? Demand is leveled by selective buffer inventories at critical links of the supply chain (RBDM).

INVENTORY MANAGEMENT

One of the critical SCRM themes in this book is to move from a pure just-in-time inventory or demand-based approach to a balanced SCRM approach. An end-product manufacturer balances and optimizes incoming, in process, and final buffer inventories.

A push, just-in-time, or other optimal supply techniques only work well when there is a steady demand for products. Supply chain spikes or troughs in demand turn into supply chain headaches. The only way to lessen their impacts is to manage inventories very carefully. Who has the inventory and who pays for it can become a shell game of risk management. Inventory may be found at the supplier's facility, in transit, in a warehouse, at a distributor, or at areas in the end-product manufacturer's facility.

Distribution centers offer a middle solution between just-in-time and buffer inventories. The distribution center can serve as a cross stock or inventory for critical products sold through retail supply chains. In some cases, containers may go directly to retail store outlets without even touching a distribution center (RBDM).

SUPPLY INVENTORY MANAGEMENT

Another solution to having sufficient parts available at critical points of the supply chain is to push inventory requirements onto suppliers. This is called supply inventory management. The end-product manufacturer and key suppliers manage parts and inventory by developing joint production schedules. Who pays for this inventory is still a critical question? The inventory is collocated with the supplier and the end-product manufacturer. Or, the supplier may carry most or all the inventory. This only works if there is mutual trust for absorbing and allocating costs. Inventory management is not the best answer to the lean supply chain. But sometimes, it is the optimal response for keeping the chain running smoothly.

Another solution is to calculate upper and lower inventory targets within the supply chain and across product categories. So if there are variances in demand, inventory can fill in for spikes in demand or can smooth out production slowdowns.   The critical question then becomes where and how much inventory to hold across the manufacturing facilities, assembly plants, packaging facilities, and local warehouses.

WALMART ON DEMAND DELIVERY RISK

On time delivery is a risk factor for suppliers. Big box retailers especially need truck deliveries within tight periods. Why? Retail goods are delivered to specific docks based on tight delivery schedules. Too early? There is too much unsold product in incoming inventory or unsold material on the shelf. Too late? There are insufficient products to fulfill orders. Then products can be packaged or marked improperly. Or, the sequence of palettes on the truck is off. All of these options cost money and present risks.

Walmart is taking this a step further. They have developed a program called 'On-Time, In-Full' to decrease schedule or delivery risks, both too-early and too-late deliveries. Walmart wants to increase operational effectiveness by mitigating operational risks.

Walmart has been working in reducing inventory and cleaning up its 4,700 stores as back rooms became cluttered with unsold instore inventory. Walmart requires full truckload suppliers to deliver 100% of their high turn-over items such as groceries, toilet paper within 75% of the agreed upon delivery schedule.[202]

REAL-TIME INVENTORY ORDERING

'Smart' vending machines are another option and can now be found in many retail outlets and malls as the following details:

> "Some drink producers (for example, Coca-Cola with its Freestyle initiative) have created 'smart' vending machines that allow con-sumers to configure their own soda flavor at the time of consump-tion. With lots of mixture options, consumers can mix together any number of their favorite flavors in seconds. Sensors monitor levels of flavor syrups in the machines and send alerts to the supply chain, which responds swiftly to replace anything a machine is running low on. Drink producers can also track customer mixes to understand preferences and trends – insights that can be used to develop new canned sodas."[203]

JUST-IN-TIME MANAGEMENT TOOLS

The first person gets the oyster; the second person gets the shell.
Andrew Carnegie, businessperson

Lean management, quality management, benchmarking management, lo-gistics management, and just-in-time management are complementary concepts. The goal of these techniques is to reduce risk, eliminate waste, add value, and optimize the entire chain of value-added activities from un-derstanding end-product manufacturer requirements to quickly delivering a product. In this section, we refer to these practices by the more traditional just-in-time term.

ROCKS IN THE LAKE

Just-in-time (JIT) is a broad philosophy covering many operational excel-lence areas. Just-in-time broadly is the design and management of cus-tomer-supply, value-added processes to minimize cycle and lead times.

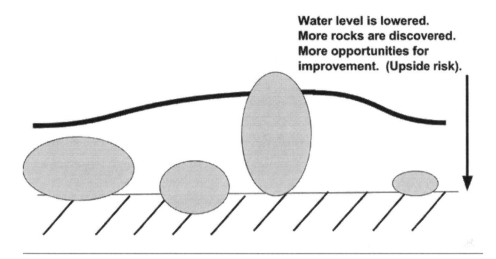

Water level is lowered.
More rocks are discovered.
More opportunities for
improvement. (Upside risk).

*Figure 38: **Rocks in the Lake***

The precise arrival of parts to the specific location is the conventional interpretation of just-in-time. However, it is but one example of the very broad impact of JIT.

JIT is often illustrated as rocks in a lake as shown in the above figure. As water is lowered or inventory is reduced, more problems are discovered which can be corrected. Inventory can hide operational problems including machine downtime, scrap, work in process, buffer inventories, engineering design redundancies, change orders, inspection backlog, paper backlog, poor material quality, high scrap rates, late supplier deliveries, transaction errors, double orders, safety stock, and rejected materials (RBPS).

SCRM ELEMENTS IN JIT
SCRM incorporates the following JIT techniques and tools, specifically:

- **Just-in-time deliveries.** Products are delivered just-in-time to be used on the manufacturing floor. Deliveries are in small quantities, supplied more frequently, and have an exact count.

- **Suppliers located close to the plant.** In order that shipments are delivered reliably, suppliers of perishable, critical, or heavy products are located near the end-product manufacturer's plant.

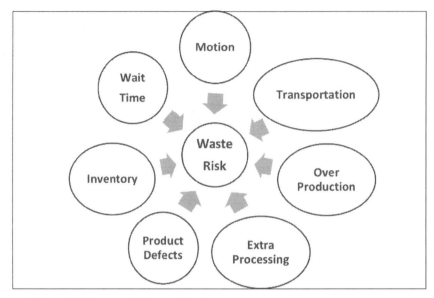

*Figure 39: **Types of Waste - Risk***

- **High-quality goods.** JIT delivery requires supplied products are defect free. If defective products end up on the manufacturer's production line then the line will stop until the defective products are replaced.

- **Reengineered internal and supply processes.** Internal processes are examined and if necessary redesigned. Streamlined processes add value and eliminate waste, which are the core values of all just-in-time processes.

- **Total quality.** Total quality in its broadest sense is the goal and objective of every value-adding, process stream. Total quality involves all elements from astonishing final-customers to ensuring products conform to specifications.

- **Controlled and capable processes.** Critical supply chain processes are stable and capable. Stable processes are in control. Capable processes meet or exceed final-customer requirements. The goal of all controlled and capable processes is to eliminate unwanted variation.

- **Customer-supply partnering.** An important element in JIT can be the need for fewer suppliers. Multiple product suppliers result in increased product, delivery, cost, and quality variation caused by lack of understanding, poor communication, or differences in capabilities. Whatever the cause, the results are the same - poor quality products and services result in dissatisfied customers.

- **Workers and management commit to work together.** The entire JIT process from order taking to product/service delivery is streamlined. Many disparate parts of the end-product manufacturer must work together as smoothly as an expensive Swiss watch. If there are problems, they are root-cause solved and eliminated (RBPS).

- **Dock to production delivery.** Dock-to-stock was the prevailing JIT wisdom for many years. Now, end-product manufacturers want to eliminate or at least minimize incoming, buffer, and final inventories and move incoming parts directly onto the production line or to the point of sale (RBPS).

- **Sequenced delivery.** The right parts have to arrive at the right location in the proper amount, at the right time, and in the right order. Even the packing of the parts in a truck or railcar is critical. For example, parts are packed in the truck in the reverse order in which they will be unloaded. To minimize handling and storage, parts are pulled from the truck in the sequence they will be used on the assembly line. Benefits from sequenced delivery include: Trucks are unloaded quickly. Parts go directly onto the production line. Trucks do not have to wait to unload. Costs are reduced. Queues are eliminated. Space is not required to repalettize or shift parts. Loading dock efficiency is increased.

- **Proper packaging.** Parts are packaged suitably for use. If parts are packaged loosely then products are damaged. If parts are packaged too robustly then additional personnel, equipment and time are required to break down the packaging.

- **Accurate demand forecasting.** Forecasting is one of the most critical activities in SCRM. Without accurate forecasts, supply chain partners can only react to orders as they are received. With constantly increasing pressure to reduce lead times, this means large

---

CONTEXT: The Seven Deadly Wastes Risks

- **Motion.** Incorrect layout of office and factory, lack of proximity of machines, off-line resources.
- **Wait Time.** Lack of coordination, idle operators watching a machine, long set-ups.
- **Overproduction.** Inaccurate forecasts, large batches, full utilization of machines and labor.
- **Extra processing.** Poor machine maintenance, unnecessary processing steps.
- **Product defects.** Rework, troubleshooting delays, dissatisfied downstream customers.
- **Inventory.** Space requirements, obsolescence, clutter, lack of forecast accuracy.
- **Transportation.** Unnecessary movement of material, extra handling.

---

inventories at various points along the supply chain. With good demand forecasts, shared with supply-partners, 'just in case' inventories are significantly reduced.

- **Accurate price forecasting.** Operations managers must also develop supply availability and price forecasts. Frequently, time and effort are devoted to developing demand forecasts, but forecasts of supply capabilities to keep up with increasing orders are often overlooked. Without them, however, parts may not arrive in sufficient quantities. Similarly, as supply segments approach capacity, prices are likely to escalate.

JUST-IN-TIME SUPPLY MANAGEMENT

Just-in-time assumes a smooth process flow of products. Let us say that a supplier gets an order of 100 parts one day and the next day it increases to 200 or even more. How is this order going to be fulfilled? Is the supplier going to work two shifts to fulfill the order? Or, is the supplier going to pull products from inventory? Both responses violate several principles of lean and just-in-time management.

One option to this significant problem is to control variation through long-term sourcing and partnering relationships. The number of suppliers may be reduced and streamlined so that some may be single-suppliers while others are part of a select group of commodity suppliers. Purchase agreements may last for years. Strong bonds develop between the manufacturer and supply-partner with each depending on the other for profitability.

JUST-IN-TIME AND SCRM METRICS/OBJECTIVES

End-product manufacturers, such as auto manufacturers have thousands of suppliers. Stratified parts, such as a component in a sub-assembly, which is part of an assembly fitting into a system may involve hundreds of parts and hundreds of suppliers. As well, many thousands of parts come from multiple suppliers from different countries that may cross many borders several times. So, it is critical that final-customer value attributes are captured by the end-product manufacturer and are incorporated into engineering drawings, specifications, and standards to suppliers. SCRM value attributes in a JIT environment may include

- Timeliness.

- Reliability.

- Product integrity.

- Customer service.

- Accurate and complete information.

- Flexibility.

Timeliness

Timeliness is the ability to deliver a product when and where the final-customer wants it. Cycle time management and speed are the keys to JIT delivery. Especially in JIT distribution, the internal customer wants products delivered onto the production line at precise schedules.

Reliability

Reliability is the ability to deliver products at scheduled intervals. Delivery reliability is measured in terms of transit time, which is compared to promised delivery. Supply management can contract for supplies to be drop shipped on a monthly basis to a specific location. The manufacturer relies on the supplier to meet delivery obligations every month; otherwise a missed shipment shuts the production line. If missed shipments become a habit, then the end-product manufacturer is required to carry additional inventory to meet unexpected demands. This is expensive because of high inventory, storage, and handling costs.

Many military systems are now aging and not reliable. For example, It is not uncommon to find 20 to 30 year-old IT systems that need to be updated, patched, or replaced. However, the original suppliers may have gone out of business, moved, or transferred their technology offshore. Each of these can result in reliability problems especially with highly complex designs. More end-product manufacturers are paying suppliers to stay in business because having a critical system malfunction may involve safety or national security (RBPS).

Product Integrity
Product integrity means a defect-free product is delivered intact to the end-product manufacturer. There is no storage, shipping, or handling damage. Once delivered, the product functions as the end-product manufacturer expects.

Customer Service
Customer service keeps the customer advised of delivery problems, providing notice of product changes and price changes, providing accurate invoicing, satisfying warranty claims, reconciling billing differences, and supplying technical assistance (RBPS).

The sales engineer provides current and accurate information of a product's special features, maintenance history, and performance levels. An order desk clerk provides prompt and efficient order taking. If product integrity has been compromised, then problems need to be resolved courteously, quickly, and satisfactorily (RBPS).

Accurate And Complete Information
Product delivery is not complete if support documentation is not current, accurate, and complete. This is especially important for a complex industrial product. Support documentation includes product certifications, parts

lists, engineering prints, spares list, operating instructions, maintenance instructions, inspection reports, and other product information.

Flexibility

Flexibility is the ability to respond to sudden customer and end-product manufacturer needs. For example, an end-product manufacturer requires a rush delivery to satisfy unexpected demands. Sudden needs may require substituting material, rescheduling shipments, or changing carriers.

JIT ADVANTAGES

JIT management is a form of risk management and SCRM. JIT is different than traditional distribution and inventory management. In JIT, material distribution, transportation, storage, and handling are synchronized to ensure reliable and stable product delivery. End-product manufacturer inventory is reduced drastically and is hopefully eliminated. Product warehousing requirements are also reduced.

Specifically, JIT risk management results in:

- Fewer suppliers.

- Improved product quality.

- Reduced inventory.

- Fewer nonproductive personnel.

- Lower costs.

Fewer Suppliers

JIT requires fewer suppliers. JIT requirements ultimately reduce the supply list to those that can comply with contractual commitments. Compliant suppliers are offered inducements, such as predictable orders for their products. Suppliers of critical products are often the most compliant, willing to learn, and try new techniques. These suppliers have major accounts and have the largest incentive to adopt new management practices.

Improved Product Quality

Quality, as 100% conforming material shipments, is required for the proper functioning of JIT. The end-product manufacturer does not inspect any incoming material shipments. If a supplier delivers a shipment with defective products, then the end-product manufacturer's production line stops. Other production lines down the supply chain also stop. This is expensive. Once suppliers understand the consequences of failure, they must deliver on time, defect-free material.

Reduced Inventory

The goal of JIT management is to progressively reduce all types of inventory. Traditionally, by maintaining high inventory levels, a supplier could provide a high level of customer service by supplying products on-demand. Inventory was a buffer to balance unpredictable supply and demand levels.

Fewer Nonproductive Personnel

As inventories are reduced, material handlers, parts inspectors, supervisors, and other employees, who do not add real supply chain value are reduced.

Lower Costs

The SCRM goal is to lower overall costs. JIT management, regardless of what it is called works because it impacts the bottom-line. Nissan Motor estimated that converting to a just-in-time integrated supply chain saves up to $3,600 per vehicle. This is more than the net profit realized from each vehicle.[204]

JIT SUCCESS

JIT drastically reduces investment as well as total supply chain costs. The U.S. automobile industry quickly realized the advantages of JIT. In the U.S., automakers were carrying $775 worth of work-in-process inventory for each car they built, while the Japanese carried only $150. The very existence of the U.S. auto industry depended on adopting the JIT philosophy. We are now in the same predicament as end-product manufacturers are often fat with high inventories and only reasonable quality.

Delphi Technologies, a GM supplier, is another JIT success story. It used lean and JIT management to increase business performance. Productivity was up 200% in one facility, lead time for deliveries was reduced by 50%, late shipments at premium rates were down 15%, and inventory turn improved by 170%.[205]

LEAN MANAGEMENT TOOLS

Nothing is more satisfying when timing and delivery occur in perfect sequence.
Anonymous

Lean management or simply 'lean' is another set of risk management tools.

WHAT IS LEAN?

The question is what is lean and how is it be implemented within the supply chain? Lean incorporates many SCRM and JIT management ideas such as sequenced delivery, lower inventory levels, and quick changing machine set up-times.

The following are common definitions of 'lean':

"a philosophy of manufacturing that focuses on delivering the highest value product at the lowest cost on time."[206]

"a systematic approach to identifying and eliminating waste (non-value-added activities) through continuous improvement ...."[207]

Many SCRM ideas are evolving into a philosophy of thinking lean and working lean. Lean first focused on manufacturing and the concept is now morphing to include the entire value stream consisting of all steps needed to convert resources into products or services the final-customer wants. Any process step that costs too much, takes too long, or does not optimize value is wasteful, risky, and is eliminated.

STARTED WITH TOYOTA

Many just-in-time and lean ideas originated at Toyota Motor Company. The principal idea was to foster flexible, low-cost, and shorter production runs

---

CONTEXT: Lockheed Principles of Lean Manufacturing

- **Visual transparency.** If you cannot see it, you cannot manage it. Visual management implies there is a clear display of charts, lists and tools.
- **Design For Manufacturing and Assembly.** DFM/A ensures products are produced easily and consistently with high-value add.
- **Process focus.** Process focus is the essential element of all supply chain, innovation, Six Sigma, and lean manufacturing initiatives. The supply chain looks at the overall process, while lean manufacturing looks at maximizing sub-processes, production cells, and machines. Each discrete operation is analyzed and stabilized before the entire process chain can be stabilized. The end-result is a streamlined flow of products.
- **Just-in-time.** In a JIT, pull system, specific products are produced as they are needed, when they are needed and only in quantities that are needed. Constraints or stresses on the weak supply chain links are quickly uncovered or discovered.
- **Process control.** Process control ensures that unusual process conditions are detected quickly, corrected, and prevented from recurring.
- **Standard work.** Fundamental to all lean is that repetitive activities are simplified, standardized and proceduralized. This means that work steps are flowcharted and procedures are written capturing best practices and lessons learned. This way if critical people leave, core processes are still stable.

---

that could meet end-product manufacturer requirements for high-quality, low-cost products. Now take this idea and apply it throughout the supply chain. This is the intent of lean supply management. Does it work? It is more difficult that it looks because design, production, and parts ordering must be seamless and smooth.

Let us look at lean initiatives at Boeing, Daimler-Benz, and Johnson Controls. When reading these stories notice how lean integrates quality, lean, JIT, cycle time, and other management philosophies.

BOEING LEAN

System sourcing (systems integrator) is a strategy where one supplier provides a major assembly as opposed to many smaller suppliers providing subassemblies or even components. For example, the Boeing Commercial Airplane group in Seattle previously produced the door liner for the Boeing 777 airplane in-house. The bill of materials for the end item consisted of more than 150 part numbers from more than 50 suppliers. The parts were ordered from different suppliers and were delivered to the Boeing 777 production line to meet Boeing's final assembly schedule. This was complicated and resulted in risks and variation for Boeing.

Lean management attempts to eliminate unnecessary variation. Boeing wanted to become lean by reducing the number of production parts suppliers. How was this done? Now, a systems supplier provides the final subassembly of the door liners. Once the final subassembly is completed, the designated supplier delivers the door liner just-in-time to be fit on the 777.[208] Under this approach, everyone wins. Boeing shifted responsibility for the subassembly, quality, and delivery to the supplier. The Boeing 777 production line now has fewer parts.

Boeing also realized the following benefits:

- Simplified bills of material.

- Reduced legal and contracting costs.

- Reduced supplier base.

- Reduced assembly costs.

- Reduced travel time and expenses by eliminating travel to numerous suppliers.

- Reduced product nonconformances.

- Reduced cycle time.

MERCEDES BENZ LEAN

Mercedes Benz assembles its new M-Class sports utility vehicle at the Vance, Alabama plant. The plant has adopted a strategy in which a single-supplier provides entire systems on a JIT basis. For example, instead of

---

CONTEXT: Lean Work Tips

- Work with selected supply-partners to help them develop lean processes.
- Level production schedules to avoid big spikes in demand, which allow suppliers to minimize inventories.
- Create a disciplined system of time periods when parts shipments have to be delivered.
- Develop lean transportation systems to handle mixed load, small lot deliveries.
- Encourage suppliers to ship what is needed to the assembly plant at a particular time.

---

buying head rests and seat cushions from different suppliers for subassembly in-house, Mercedes receives fully assembled seats from a designated supplier

There are more than 200 deliveries a day to the Alabama plant, from more than 65 sources. Suppliers are required to sequence their deliveries so they arrive at the plant in the proper order for daily production.

JOHNSON CONTROLS LEAN

In many ways, lean management is similar to just-in-time management. For example, lean delivery involves a just-in-time relationship with suppliers. In one case, Toyota formed a close partnership with Johnson Controls to deliver seats to be just-in-time installed on the assembly line. Inventory levels dropped form 32 days of inventory to 4.1 days. Along with this, set-up times for dies and machinery were reduced from hours to as little as 17 minutes.[209]

QUALITY MANAGEMENT TOOLS

One consequence of postwar technology has been the acceleration of change in our society, so that we seem to produce a new generation of products about every five years.
Ross Macdonald

Quality management/assurance/control are another set of commonly used risk tools. High-quality products and services are essential to supply chain management. If a supplier produces nonconforming products, these rejected products can cause production lines to stall unless there is some buffer inventory.

Today's approach to quality is called Six Sigma. Six Sigma companies call it a philosophy, a set of guiding principles as well as a set of tools. Regardless of the term, Six Sigma is a widespread method for improving production quality and delivery of goods and services.

THE SIX SIGMA QUALITY REVOLUTION

Quality like purchasing has gone through several changes. Quality management has been called process innovation, quality control, total quality management, business process innovation, and now Six Sigma. Are these the same? No. However, the intent is the same, which is to satisfy final-customers with improving products delivered on time and on budget. More often, the supplier intent is to exceed end-product manufacturer expectations with improved products and services. In commercial and industrial supply management, the end-product manufacturer wants products delivered in sequence in tighter time windows while lowering overall costs.

Six Sigma has almost become a cult of perfectibility. GE's CEO Jack Welch single-handed launched Six Sigma. Before that, it was a techie statistical tool for improving a manufacturing process. Welch needed a measurable methodology to baseline and benchmark business performance. Six Sigma fit the bill perfectly. All of a sudden, it became an enterprise innovation ethic. And it worked. According to GE, Six Sigma added $600 million to GE's bottom-line.[210]

PARTS PER MILLION QUALITY LEVELS

End-product manufacturers in competitive environments expect parts per million (PPM) quality levels from suppliers. Historically, firms purchased parts according to Acceptable Quality Level (AQL) criteria. Using AQLs, companies usually accepted products with 1,000 or 10,000 parts per million defect levels.

SCRM processes require very high consistency and high-quality levels. Six Sigma quality by definition is 3.4 parts per million defect levels. Very high-quality? You bet. Thousands of U.S. companies are pursuing Six Sigma.

CONTEXT: Mikel Harry's Six Sigma Methodology

Measure:
1. Select CTQ (critical-to-quality) characteristic.
2. Define performance standards.
3. Validate measurement system.

Analyze:
4. Establish product capability.
5. Define performance objectives.
6. Identify variation sources.

Improve:
7. Screen potential causes.
8. Discover variable relationships.
9. Establish operating tolerances.

Control:
10. Validate measurement system.
11. Determine process capability.
12. Implement process controls.

Fad du jour? Maybe! The reality is that many U.S. companies still have quality levels around 3000 parts per million.

RELIABLE COMPONENTS

A finished product is only as reliable as its smallest component. Since a finished product is the sum of many small parts, which form a subassembly, assembly, and finally, a finished product, the smallest component can cause the whole unit to fail. For example, if a tiny rivet on an aircraft bulkhead fails, there is a cascading effect where the pressurized bulkhead buckles and causes the pilot to lose control of the aircraft. If a manufacturer obtains many small components from suppliers, each component has to be as robust and reliable as the finished product.

In electronics, parts per thousand defect rates especially cannot be tolerated. For example, many electronic components are wired in series. If one series component fails then the whole component can fail. In a hypothetical

---

CONTEXT: The Motorola Story

Motorola is among the first companies to promote Six Sigma techniques. Now Motorola and others use these systems to drive higher levels of innovation with suppliers. Motorola expects verifiable innovation in four critical areas:

- Keeping pace in attaining perfect product quality.
- Remaining on the leading edge of product and process technology.
- Practicing just-in-time manufacturing and delivery.
- Offering cost-competitive service.

---

computer that had 10 printed circuits boards each containing 10 components, if the parts were 1% defective, then as many as 97% of the computers would end up with at least one defective part. Depending on the wiring configuration, one defective component could create a malfunction making the whole computer inoperable. The solution: electronic operations managers are requiring suppliers to ship parts with no more than 10 parts per million defect rate.

WHY DOES SIX SIGMA WORK?
Why has Six Sigma worked when other quality initiatives have stalled? Six Sigma works. Money is saved. It is relatively simple to apply. Employees get trained in problem solving (RBPS) as 'project champions' or as 'black belts.'

Six Sigma projects are usually doable, measurable, and manageable. This ensures demonstrable business results. While results can vary, innovations of $50K to $250K in cost savings or cash generating impacts are often common.

BENCHMARKING MANAGEMENT TOOLS

If we have had a formula for growth it has been; start with the best; learn from the best, expand slowly and solidify our position; then horizontally diversify our expertise.
Mark McCormack, writer

Benchmarking is a critical method for measuring SCRM innovation. Benchmarking is the continuous process of comparing and measuring processes, systems, services, practices, and products against leading companies inside or outside one's industry sector.

'BEST IN CLASS'

Benchmarking has been around for about 20 years. It is pretty simple. It looks at who is doing what, usually called best practices and then compares these against what is done internally. A gap analysis then reveals what best practices are integrated internally or with suppliers.

Often a benchmarking study is the impetus for an end-product manufacturer to discover that something is wrong and the solution is to adopt SCRM. The competition is doing something critical in half the time and is reaping tremendous profits from this SCRM strategy. Evaluating and if necessary redesigning internal supply chain processes is the first step to increasing time-to-market. First a company eliminates high risk, wasteful, or nonproductive internal tasks and then external supply chain processes are targeted for similar innovation.

End-product manufacturers can benchmark just about any activity. For example, the following SCRM practices can be benchmarked: electronic data interchange (EDI), supply certification, quality practices, project delivery, commodity costs, manufacturing resource planning (MRP), JIT delivery, and customer-supply partnering

TYPES OF SCRM BENCHMARKING PROJECTS

There are 4 basic types of benchmarking projects: 1. Internal, 2. Competitive, 3. Functional, or 4. Generic benchmarking.[211] Often, benchmarking companies look for tips outside an industry sector because it is easier to gather information. These companies are not paranoid about a competitor stealing a secret and are more willing to share information.

SCRM benchmarking is a one-time effort or it is a continuous process of comparing an end-product manufacturer's SCRM against a competitor's. End-product manufacturers that continuously benchmark learn to adapt and adopt new SCRM practices quickly. It is critical that benchmarking end-product manufacturers look for companies that have 'world-class' SCRM processes. A benchmarking project as any related SCRM initiative

---

CONTEXT: SCRM Benchmarking Warning Signs

- Lack of executive management sponsorship.
- Wrong people on the SCRM benchmarking team.
- Lack of true understanding of how the end-product manufacturer deployed SCRM practices.
- Unmanageable SCRM benchmarking team.
- Underestimation of time, resources, and efforts required to complete and implement benchmarking results.
- Over emphasis on reaching SCRM performance targets instead of focusing on innovation processes.
- Use of benchmarking as a tool, instead of a SCRM positioning strategy.
- Use of benchmarking for minor challenges instead of furthering SCRM strategic objectives.
- Too many site visits especially when information is researched or is a phone call away.
- Failure to follow up on SCRM benchmarking implementation.

---

reinforces the end-product manufacturer's strategic SCRM vision and mission.

Unfortunately, the above requirements are not always followed. The scope of the SCRM benchmarking project may be too broad and not easily achievable. For example, a study to benchmark the best practices of a wafer fab operation is difficult if not impossible. These plants may cost a billion dollars. They have state-of-the-art technologies that end-product manufacturers do not want to share with competitors.

In general, benchmarking properly deployed offers the following benefits:

- Establishes achievable innovation targets.

- Breaks down the 'why break it if it works' thinking.

- Destroys preconceptions.

- Initiates an organizational and supply chain cultural change.

- Establishes a SCRM innovation methodology.

- Sets accountabilities for supply process and product innovation.

HOW TO BENCHMARK SCRM PROGRAMS

Robert C. Camp, the author of the best-selling **Benchmarking**, outlined the following steps of a successful benchmarking project:

- **Identify what is to be benchmarked.** Define the mission, deliverables, and performance measurements of the SCRM benchmarking project. Understand the SCRM process, product, or procedure to be benchmarked.

- **Identify comparative supply chains.** Identify the best SCRM competitors or industry leaders from whom supply chain risk lessons can be learned. As well, determine the appropriate form of risk benchmarking, i.e. competitive, internal, functional, or generic benchmarking. Then, approach selected companies to be benchmarked and have backup companies. Identify constraints to benchmarking within or outside one's industry.

- **Determine a data collection method and then collect the data.** Determine how SCRM information is to be gathered. SCRM information is collected from internal sources such as internal experts or through public domain information such as external experts and consultants.

- **Determine current performance 'gap'.** Gap analysis investigates differences in present risk-control practices, performance, cost, quality, or efficiency against those that were benchmarked.

- **Project future performance levels.** Is the SCRM gap widening or closing and at what rate? This analysis provides understanding of the gap and what can be done to close it by deploying new risk-control practices or procedures.

- **Communicate benchmark findings and gain acceptance.** The results of the SCRM benchmarking study are communicated to the appropriate stakeholders by providing specific recommended actions.

- **Establish functional goals.** SCRM benchmarking results are then operationalized, specifically translated into functional, attainable goals.

- **Develop action plans.** The actions to achieve SCRM functional plans and goals may involve process controls, establishing process capability, developing new training methods, or pursuing other system/process innovations.

- **Implement specific actions and monitor progress.** Specific actions are deployed to achieve SCRM goals. Implementation is through line management or supply management. Progress is then continuously monitored.

- **Recalibrate benchmarks.** Progress reports determine if SCRM benchmarked practices are implemented according to plan. If not, the process is corrected or recalibrated.[212]

LOGISTICS MANAGEMENT TOOLS

You will not find it difficult to prove that battles, campaigns, and even wars have been won or lost primarily because of logistics.
Dwight D. Eisenhower

Logistics is the movement of goods and parts within the supply chain. Traditionally, logistics was a risk or weak link in the supply chain because truck, plane, and ocean carriers could not identify and locate goods. Suppliers would carry buffer or safety inventory so a missed or delayed shipment would not stop production. Well, technology has made this much easier with real-time GPS monitoring and real-time traceability with RFID tags.

WHAT IS LOGISTICS MANAGEMENT?

Logistics, simply defined, is a set of practices to determine how to move people and materials most efficiently between a given source and destination. If logistics is not economic, efficient, or effective, then risks can arise. The 'supply chain' metaphor further extends this idea to denote a group of loosely connected companies, all collaborating on the efficient and economic delivery of products.[213]

APICS has a more formal definition of a logistics system:

"The planning and coordination of the physical movement aspects of a firm's operations, such that the flow of raw materials, parts, and finished goods is achieved in a manner that minimizes total costs for the levels of service desired."[214]

The critical point is that a supply chain is only as good as its product delivery system. The movement, preservation, packaging, and inventorying of products are managed throughout the supply process from critical supply-partners to end-product manufacturers to users. Again, the goal is to reduce variation, reduce risk, and ensure consistency. Full truckloads of the correct materials packaged suitably are sent to the right location, in the right sequence for unloading at the right time. Less than truckload costs are also managed. Truckload times are reduced. Products are bar coded. Product counts and inspection are eliminated entirely.

Logistics management has many common elements with just-in-time management. For example, critical elements of logistics management include:

- **Elimination of incoming receiving inspection.** Inspection is entirely eliminated. The quality engineer may receive an incoming Statistical Process Control (SPC) chart indicating incoming products were process-controlled. Quality engineers develop quality standards to which the suppliers must demonstrate compliance.

- **Direct logistics.** Materials come directly from the supplier. Repeated transferring among different transporters is eliminated thereby reducing transportation damage.

- **Accurate counts.** Quantities delivered to the end-product manufacturer are exact. Shipments are delivered on an as required basis. There is no permanent incoming or buffer inventory. Material arrives just-in-time to be used.

LOGISTICS MANAGEMENT
Trucks, trains, ships, and planes carry products. The mode of transportation is determined by the value of the goods, cost of transportation, and

demand for the products. The operations manager often determines the best method of transport.

As cycle times shrink and demand for goods changes, operations managers find their work is becoming more difficult. Manufacturers will not stockpile parts wanting them just-in-time for use. So, the responsibility for balancing transport flow and ensuring timely delivery falls on the operations manager.

To ensure JIT delivery, operations managers and logistics experts want real-time, status information of the shipment. How is this done? The buyer and carrier are electronically linked. The carrier has a wide array of technologies that can be used. Shipments and products are bar coded, tracked using GPS, authenticated using block chain, identified using RFID tags, and assigned to a specific carrier. Carriers then monitor the shipment using satellite tracking technology.

SUMMARY
SCRM labels are damaging. Let us look at a few. If we need to be lean, what were we before, fat and lazy? If we need Six Sigma quality, what were we before, shoddy and sloppy? If we were doing just-in-time, were we doing just late?

Supply chain management is based on simple management truisms, such as implementing lean, reducing variability, managing by exception, and managing using best practices. Lean implies cutting all fat from all processes. Supply stream implies the entire production or service chain is smooth. While the techniques discussed in this chapter are obvious, they are not consistently applied. Why? They take energy, daily commitment and challenge the ways people work.

NEXT CHAPTER
In the next chapter, we discuss SCRM emerging topics.

CHAPTER 15:
SCRM EMERGING TOPICS

WHAT IS THE KEY IDEA IN THIS CHAPTER?
Emerging topics such as sustainability, cyber security, and government contracting are adding more complexity to the SCRM space. Over the next 5 years, each of these will become more prominent. National governments are adding sustainability reporting requirements into statutes. More supply chains are being hacked to secure confidential and proprietary information. Governments are adding new sourcing requirements due to SCRM risk. In this chapter, we discuss these SCRM emerging topics.

SUSTAINABILITY PRACTICES
We care about every worker in our worldwide supply chain … What we will not do – and never have done – is stand still or turn a blind eye to problems in our supply chain. On this you have my word.
Tim Cook, CEO of Apple

Both the private and public sectors are adding sustainability practices to SCRM. Many public sector and national governments are also placing sustainability requirements into statutes and rules, while the private sector is developing SCRM best practices.

SCRM SUSTAINABILITY
Apple, Nestle, Walmart, McDonalds, and other companies are developing supply chain sustainability policies to conserve natural resources, improve reputations, and reduce carbon emissions. This often means there is a shift in SCRM practices. For example, companies work harder to choose suppliers based on locations where wages, working conditions, safety, ethics, and other social equity issues are similar to the host company.

CONTEXT: Walmart Sustainability Standard for Suppliers

- **Compliance with laws.** Comply with national and/or local laws.
- **Voluntary labor.** Labor is voluntary, including no slave, child, under-age, bonded, or indentured labor.
- **Labor hours.** Suppliers provide workers with rest days.
- **Hiring and employment practices.** Suppliers verify workers' ages and legal right to work.
- **Compensation.** Suppliers compensate workers with wages, overtime, and benefits.
- **Freedom of association and collective bargaining.** Workers can choose to join trade unions and bargain collectively.
- **Health and safety.** Safe work environment is maintained.
- **Dormitories and canteen.** Safe, healthy, and sanitary factory living and eating facilities are designed.
- **Environment.** Factories comply with environmental laws including waste disposal, air emissions, discharges, toxic substances, and hazardous waste.
- **Gifts and entertainments.** Suppliers do not offer gifts or entertainment to Walmart employees.
- **Conflicts of interest.** Suppliers do not enter into relationships that create conflicts of interest with Walmart.
- **Anti-Corruption.** Suppliers do not tolerate, permit, or engage in bribery.
- **Financial integrity.** Suppliers maintain accurate records based on standard accounting practices.[215]

Corporate Social Responsibility and sustainability are critical concerns that Boards must address. Forward looking, end-product manufacturers look for ways how the company can embed sustainability into the organization's strategy and ethos.

Many global companies require Corporate Social Responsibility (CSR) standards of their suppliers. For example, Walmart's standard is typical of CSR standards:

"The safety and well-being of workers across our supply chain is important to Walmart. Our Standards for Suppliers, along with our Standards for Suppliers' Manual, make clear our fundamental expectations for suppliers and factories. All suppliers and their facilities – including subcontracting and packaging facilities – are expected to uphold these standards."[216]

Potential problems with a small supplier far removed from a large end-product manufacturer can endanger a multinational company's reputation through negative social media and general business media.

SUSTAINABILITY STATUTES

France recently drafted a supply chain responsibility law. The law requires French companies to manage environmental risks in their global supply chains. French companies have to be vigilant and develop due diligence processes to identify risks to human rights, institute better working conditions, and enhance the environment in their supply chains. These companies have to identify areas of risk and carryout regular audits of their suppliers and contractors. If there are noncompliance issues, companies must put in place measures to reduce risks and develop mechanisms to monitor risks.[217]

WATER RISK

The importance of sustainability can be seen in the recognition that climate change can result in unforeseen consequences. Let us look at water and agriculture. Agriculture is a leading source of pollution. It is estimated that one-third of the world's food is grown in areas of high water stress. The food sector uses up to 70% of the global freshwater supplies, primarily for growing crops. Freshwater management is now coupled with sustainable food management.[218]

Water risk also impacts the food and beverage sector. This sector is huge. The sector consumes 10% of global consumer spending and 40% of total employment worldwide. Fresh water is a critical element of the spending, consumption, and employment in the food and beverage sector. And, this sector is expanding so quickly that by 2030, 40% of global water demand will not be met.

Food and water scarcity is anticipated to create political instability and supply chain disruption such as cost increases, growth constraints, and changing demand patterns.[219] National regulations are also imposed on this sector for product packaging, production risk-control requirements, safety, carbon, and social reporting.

SUPPLY CHAIN SLAVERY RISK

Human rights issues are part of the organization's Corporate Social Responsibility (CSR). The challenge for a global organization is a 4th or 5th tier supplier may use child labor but the end-product manufacturer or brand owner does not know this is occurring.

The size of human slavery is difficult to estimate globally since much of it is hidden. Human trafficking, slavery, people trapped in boats, child labor, prison labor, poor wages, and inhospitable living conditions are part of human slavery. And, it is big business. Profits from modern slavery are estimated to be around $150 billion.[220] This is both a social issue as well as an equity issue. More end-product manufacturers are aware of this because of their desire of being corporate stewards and complying with Corporate Social Responsibility statutes. Many global companies are discovering that being good corporate citizens can pay dividends.

SUPPLY CYBER MANAGEMENT

It takes 20 years to build a reputation and five minutes to ruin it.   If you think about that, you'll do things differently.
Warren Buffett, CEO of Berkshire Hathaway

Global end-product manufacturers with thousands of suppliers, partners, and other stakeholders face cyber security risks. Outsourcing, global sourcing, digitization, IoT products, new sourcing models, and complex products have created new cyber challenges and risks. Reporting these risks is now a Board and executive level issue. Bad software and system vulnerabilities are now a growing source of concern.

DAILY RISK CHALLENGE

Let us assume that you are an executive of an end-product manufacturer with highly proprietary software that sells for high margins. Or, you are a trusted, high end retailer, that sells its products at high prices. The high market capitalization of both companies is largely validated by its reputation of great management with reliable risk-controls.

Except one day, worst-case scenario occurs. Proprietary data, personal data, and Intellectual Property are stolen through a cyber security breach. The company's unique value proposition may vaporize since competitors now have access to the company's proprietary data. Personal data of millions of credit card holders is now in the hands of bad people, who can destroy the trusted position of the company. And of course, the theft of data is now on the front page revealing the lack of adequate management risk-controls. The company's reputation is now lost. But most importantly, Wall Street's confidence in the organization's management plummets.

SOFTWARE RISKS

Purchased or third-party software is a growing area of concern. Third-party software is often purchased off the shelf and may not be extensively tested prior to use. The normal software verification and validation testing may not be done making it vulnerable to a beach. The result is malicious software programs or apps can creep into supply chains and result in multiple SCRM problems.

Over the last 10 years, more products have built in intelligence and accessibility. This software is a critical part of the Internet of Things. These products have an IP address and are known to be vulnerable. The software or source code can be from an unknown foreign source, which can present additional critical risks.

SCRM has become critical throughout the procurement process for every Federal agency involved in the design, acquisition, delivery, deployment, and maintenance of military products and services. Mission-critical products, materials, services that have a military or homeland security application may not have gone through a rigorous risk examination involving the identification, assessment, and remediation of vulnerabilities.

CONTEXT: ISO 28001 Supply Chain Security Standard

- General.
- Identification of the scope of security assessment.
- Conduction of the security assessment.
- Assessment personnel.
- Assessment process.
- Development of the supply chain security plan.
- Execution of the supply chain security plan.
- Documentation and monitoring of the supply chain security pro
- Continual improvement.
- Actions required after a security incident.
- Protection of the security information.[221]

IT AND CYBER SECURITY CHALLENGES

Recently, Delta Air Lines and British Airways had critical IT equipment failures. In Delta's case, 2300 flights were cancelled over three days as computer terminals were entirely shut down. In British Airway's case, an IT problem shut down the check-in system. In both cases, gate agents had to manually check-in customers. This required additional time and employees. Flights were canceled. Queues were long. Customers were unhappy. Airlines had to revise their earnings forecast (RBPS).

WHO OWNS CYBER RISK?

Cyber risk was once the purview of IT. Most companies did not have a Chief Information Security Officer (CISO). Now things have changed. Information and IP security are now Board issues. Senior executives are very concerned about cyber exposure due to cyber-attacks.

As we have discussed, software now controls much of an automobile. Automobiles are currently smart. Within a few years, software in autonomous vehicles will control all of an automobile's systems. Software risks will increase. For example, automotive manufacturers can now update software automatically. The ease by which software can be updated is a problem. BMW Board, the German automaker, announced that it would recall 11,700 luxury 5 and 7 Series models:

"In the course of internal tests, the BMW Group has discovered that a correctly developed software update was mistakenly assigned to certain unsuitable model-versions."[222]

CYBER SECURITY PROGRAM

Supplier cyber security programs are now integrated with SCRM. As this book is updated, this will become a more important chapter. While there is no way to establish an absolutely effective supply chain cyber program, the following are useful elements to consider:

- Identify critical information within the organization.

- Determine where the information resides.

- Determine the applications that process the information.

- Determine how the information is used and processed.

- Determine the vulnerabilities of the transmission and retention of the information.

- Conduct a cyber security risk assessment, evaluating threats, vulnerabilities, and controls.

FEDERAL RISK MANAGEMENT PROCESS

Long range planning does not deal with future decisions but with the future of present decisions.
Peter F. Drucker

SCRM has become critical throughout the procurement process for Federal agencies involving in IT design, acquisition, delivery, deployment in maintenance products and services. Mission critical products, materials, services that have a military or homeland security application are now required to go through rigorous risk examination involving the identification, assessment, and elimination of threats.

All U.S. federal agencies procure information systems and software. Both can be vulnerable. IT systems can be designed and developed by multiple firms including offshore firms that may have introduced malware. Software

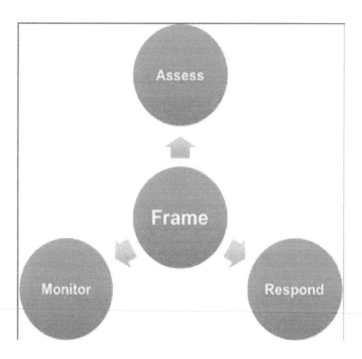

Figure 40: Federal Risk Framework

can contain malicious functionality or be vulnerable to hacking. These in-crease supply chain risks. Supply chain fraud is another source of risk. Defective products, counterfeit parts, and other risky products can be intro-duced into the supply chain.

U.S. FEDERAL CYBER SECURITY FRAMEWORK

IT and software SCRM are mandated because they can pose major vul-nerability threats. The U.S. federal government has adopted and now re-quires suppliers to comply with cyber risk management frameworks. These frameworks are increasingly important. However, the topic is so vast, that we will develop a book on this topic later in the year.

Federal cyber risk management frameworks are similar to ISO 31000 and COSO frameworks. Many recommendations in this book for architecting, designing, deploying, and assuring a SCRM initiative can be used in cyber security.

One Federal cyber security framework is composed of the following components, which are similar to SCRM (see figure on the previous page):

- **Frame risk.** Determine the context for RBPS and RBDM. This is based on the current state of the supply chain infrastructure, processes, policies. Develop SCRM policy, vision/mission, enterprise objectives, risk-control assessment methodology, and risk impact levels (RBPS).

- **Assess risk.** Determine the criticality, vulnerability, threats, impacts, likelihood, and appetite for risk. Conduct a vulnerability risk assessment and determine risk appetite (RBPS) as shown in the figure on the next page.

  The vulnerability assessment looks at potential negative impacts based upon the criticality of materials, products, and services being sourced. The risk assessment is often event based. In other words, the end-product manufacturer looks at a list of events that may pose potential risks and vulnerabilities. Risk is based on the criticality of the product, likelihood of the event occurring, and its impacts. The assessment may look at the potential harm caused by possible loss, damage, or compromise use of the supplied product, service, or material. These risk reassessments are conducted early in the supply management and acquisitions cycle to gage the overall supplier risk.

- **Respond to risk.** Design, deploy, and tailor the appropriate risk management treatment. The risk response is based on the appropriate level of control based on the supplier's context, level of control, and appetite. Response options include: avoiding, mitigating, sharing, or transferring risk (RBDM).

- **Monitor risk.** Continuously review changes of the SCRM system based on efficient and effective communications and feedback loops to ensure continuous innovation. This involves assuring that risks are controlled within the risk appetite of the organization (RBDM).[223]

C-TPAT PROGRAM
The security of shipments has been a growing concern for several years. The U.S. developed the C-TPAT program. In this program, suppliers and

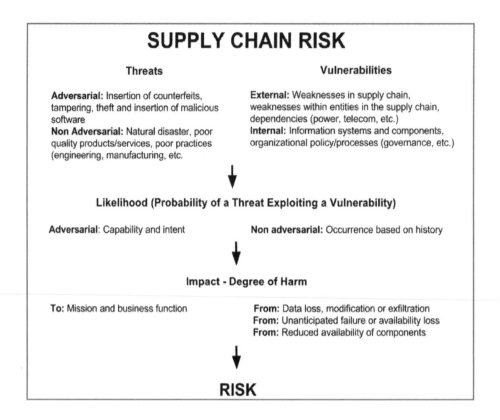

*Figure 41: **Federal Vulnerability Assessment***

shipping companies are evaluated based upon a maturity based system of risk-controls. Lowest level suppliers are certified into the program. Highest level suppliers have security processes that exceed the minimum requirements and demonstrate best practices.[224]

SUMMARY

The three topics covered in this chapter in two years will be more critical.

Sustainability practices are moving into statute. The Modern Slavery Act is a U.K. statute promulgated in 2015. The Act requires 12,000 companies operating in the U.K. to detail steps they are taking to combat slavery in their operations and supply chains. Countries are developing similar supply chain provisions requiring companies to report on their efforts to stop the use of slave labor in their supply chains. U.K. companies must also develop policies, conduct risk assessments, and monitor slavery conditions in their supply chain.

A weekly event hitting the news seems to be a cyber security breach such as a retailer's IT systems being hacked with loss of thousands or even millions of credit card numbers; ransomware attack of a hospital's information system holding its patients private information at risk; risk-control breaches in a power plant; or loss of personal data for millions of federal employees.

Department of Defense supply chains are massively distributed global networks of suppliers. Nodes in the supply chain are entry points for malicious actors. Gartner estimates that by 2020, 25 billion sensor devices will be connected to the Internet of Things. In other words each device will have a distinct IP address that can be an entry point into the network. The fear is that enemy nation states or product counterfeiters will embed sensors or malicious software in weapons systems and thus compromising military capability and safety.

NEXT CHAPTER
In the next chapter, we discuss SCRM futures.

CHAPTER 16:
SCRM FUTURES

WHAT IS THE KEY IDEA IN THIS CHAPTER?

Is SCRM here to stay? A common criticism is that every two years or so, there is a new killer fad or buzzword. SCRM is today's killer supply chain solution. Management consultants have a tendency to rediscover and re-package concepts that were introduced 10 or even 20 years ago. Will SCRM be tomorrow's fad? We do not think so. However, supply chain management does face challenges.

The bottom-line is that sourcing professionals are going be more well-rounded. This cuts to the question: 'What is the future of many supply chain and operations professionals if they do not upskill?'

SCRM CHALLENGES AND PREDICTIONS

I find the great thing in this world is, not where we stand, as in what direction we are moving.
Oliver Wendell Holmes, Writer

End-product manufacturers are becoming so efficient that Hewlett Pack-ard, IBM, Texas Instruments, and others are handing over more critical manufacturing, quality control, and distribution to suppliers.

End-product manufacturers may evolve into a core group of employees who manage a company's brand and Intellectual Property, much as we discussed in Chapter 2. This has been going on for years. The *Wall Street Journal* warned:

> "The U.S. contract manufacturers are helping hollow out Ameri-can's corporations while bolstering their manufacturing base. They

(suppliers) land orders because they are considered among the world most efficient manufacturers, and their proximity is prized because it facilitates quick product development."[225]

VUCA PREDICTIONS

Predicting the VUCA future is difficult. There are simply too many unknowns and unknowable's. VUCA factors will need to be considered in architecting, designing, deploying, and assuring a comprehensive SCRM model for the supply chain. However, the following predictions seem realistic for the next five years:

- VUCA and disruption will increase.

- New competition will arise based upon new business and SCRM models.

- 'Make or buy' decision will become more difficult.

- Product life cycles will adjust based on the perceived risk of the supplier not producing or servicing within end-product manufacturer business requirements and SCRM objectives.

- Public safety and security concerns will increase with additional cyber security sourcing mandates.

- Heightened regulatory environment will keep governance, risk management, and compliance in the forefront of all global companies.

- Outsourcing and domestic sourcing will be based on SCRM and hedging of options.

- Process of architecting, designing, deploying, and assuring SCRM across the enterprise and into supply chains will increase.

SCRM PREDICTIONS

We believe SCRM will increase in importance. SCRM architecture, design, deployment, and assurance are not easy because they involve behavioral change. The importance of SCRM will increase as a result of:

CONTEXT: Key SCRM Outcomes

- SCRM is a strategic issue. Supply chain risks are increasing and material, which require Board reporting. Supply chain risk events are becoming more costly. As a result, 71% of executives said that supply chain risk is important in strategic decision making (RBDM) at their companies.
- Margin erosion. Sudden demand changes cause the greatest impacts to the core of the business.
- Executives are concerned about risks to the extended value chain. Extended value chain includes outside suppliers, distributors, end-product manufacturers, and final-customers.
- SCRM is not always considered effective.
- End-product manufacturers face a wide variety of supply chain risk challenges including collaboration, end-to-end visibility, and justifying investment in supply chain risk programs.
- Many end-product manufacturers lack the latest SCRM tools. [226]

- Intense worldwide competitive pressures to reduce overall supply chain costs, drive to develop innovative products, and increase the focus on SCRM.

- Enhanced innovation and continuous improvement throughout the end-product manufacturer and the supply base.

- Emphasis on good governance, risk management, and compliance (GRC).

- Transportation of goods will shift depending upon the risk profile of the product and supplier.

- Domestic sourcing will become more critical as national politics impinge and even dictate sourcing options.

- Final-customers, stakeholders, end-product manufacturers, and interested parties will gain power, as information is continuously available.

- Reporting of supply chain risk and materiality issues to the Board of Directors, investment community, and regulators.

- Focus on ERM within the end-product manufacturer and into the extended organization involving suppliers.

- Emphasis on RBPS and RBDM at the end-product manufacturer and customer-supply base.

- Focus on integrating operational and supply chain core competencies with the strategic direction of the end-product manufacturer.

- Improved build to order, process capability, product quality, product delivery, cycle time reduction, service innovation, digitization, 3D printing, cloud, artificial intelligence, and ERM.

- Emphasis to provide integrated solutions that involve end-product manufacturer, suppliers, and retail shops.

- Increased use of Key Risk Indicators (KRI's).

- Standardization throughout the value chain to ensure consistency, such as simultaneous design, lean manufacturing processes, mistake proofing, total productive maintenance, and collaborative design teams.

- Streamlined and uninterrupted flow of material through the process chain to support cell manufacturing, pull/push production techniques, flexible operations, simplification, rapid machine/process changeover, and customer-supply development.

- Continuous elimination of waste, simplification, and redundancy across the value chain.

SCRM LESSONS LEARNED

Prediction is very difficult, especially about the future.
Neils Bohr, Nobel Physicist

The SCRM need is well understood, but the people, processes, and practices infrastructure are not sufficient to meet the needs. There are not enough trained operations managers to fill the demand.

FIGHTING FLAVOR OF THE MONTH

A critical challenge is that SCRM projects may fail within the next few years. Goals are not reached. Competing but complementary flavors of the month such as operations excellence, IoT management, artificial intelligence, or Six Sigma will compete for management's attention.

SCRM is too critical for competitiveness that may suffer a fad backlash. Let us look at what happened to Total Quality Management (TQM). Expectations were built up. TQM was heralded as the panacea, the miracle cure, the way to competitiveness, the holy grail of management, and so on. The problem was that there was a quick fix fixation. Reality set in. Results were less than expected. The silver bullet turned out to be a zinc pellet (RBPS).

SCRM may also become faddish. People tend to dismiss instant benefit labels. This happened with MBO, Six-Sigma, TQM, etc. All of these were useful and offered organizational benefits. Previous technologies promised, perhaps over promised 'instant pudding' benefits and failed. True reform and SCRM transformation take time, require consistent management commitment, and require increased organizational maturity and capability.

SCRM is an enterprise level, holistic, and integrated approach to performance improvement and innovation. We are seeing more executive level interest. This is critical because one critical success factor stands out in SCRM among all others - active executive management support. A journalist reported: "among people with experience, there is consensus on why supply chain management projects succeed or fail: it only works if executive management is committed and involved." [227]

HARD LESSONS LEARNED

Supply chain risk management has a long and prosperous future if developed and deployed properly. The following are my 'lessons learned' from many initiatives that became faddish and eventually failed:

- Lack of executive management commitment.

- Too much exhortation.

- Fear.

- Confusion.

- Special project mentality.

- Quick fix mentality.

- Far out technology.

- Lack of instant gratification.

Lack of Executive Management Commitment
Lack of active executive management commitment and leadership is the major cause of SCRM failure. SCRM organization and supply chain transformation are phenomenally difficult. All supply chain activities must be final-customer focused and aligned with the organizational vision.

Too Much Exhortation
W. Edwards Deming was right. It is easy to become caught up in the hoopla and excitement of a new initiative, such as SCRM. Exhortations, banners, inducements, and publicity are useful at certain stages of the process, but are counter-productive or destructive if used all the time. They lose their effect to induce, promote, or reinforce positive images of supply development and innovation.

When is exhortation useful? Exhortation is useful at the beginning of the process to create SCRM value awareness. It is useful at special occasions when the supply chain celebrates successes. It is helpful in elevating issues, awareness, and attitudes of the benefits of customer satisfaction, supplier benefits, and continuous innovation throughout the supply chain.

Fear
Fear of change paralyzes people and suppliers. SCRM is adopted by middle purchasing management who cannot move from a transaction focus. Suppliers cannot or will not adopt the SCRM even though the consequence

may be elimination from the approved supply bidder's list. Or, internal employees may not adopt process management, lean, or Six Sigma because they are associated with downsizing. Who wants to see his or her job disappear?

Confusion
SCRM may seem difficult because it encompasses a hodgepodge of ERM management principles, techniques, and skills. SCRM for many traditional purchasing managers does not make sense because it integrates benchmarking, inventory management, Six Sigma, and other practices.

Special Project Mentality
SCRM may seem to purchasing, functional, and operating managers as another special project on top of the daily concerns of moving products out the door on time. A middle manager may have to change the way he or she does a job. This is difficult and is resisted unless a good business case can be made to make his or her life easier. Change without a good rationale is always resisted.

Quick Fix Mentality
Risk-control to manage a supply chain seems straight forward. A consulting company may say: "Our enterprise resource planning, customer resource management, or data mining tools will help you do the job." The operative word is 'help' you do the job.

The reality is that supply process owners are still responsible for meeting SCRM objectives; developing risk-controls and mitigating risk; and getting the right products and services to the right location on time and on budget. There is no quick fix.

Far Out Technology
Sophisticated information technology, IoT, and artificial intelligence tools allow an end-product manufacturer to model supply processes; mine his-

torical data on product nonconformances, late/early deliveries, cost over-runs, low product counts; then identify predictive benchmarks, which are fed into the model to monitor, control and manage the supply chain.

This sounds great on paper but reality bites. There are too many suppliers, processes, people, resources, products, and unknowns that make accurate and reliable forecasting difficult. Risk-controls are difficult to apply. Most supply chains are nonlinear and often unpredictable.

Supply chain risk management sounds great. However, it is extremely difficult. There is no set of rules for a successful SCRM initiative. For example, just-in-time, demand-pull production makes sense for all parties. However, delivery and production curtailment risks are real in tightly integrated BTO supply chains. One hiccup, labor dispute, broken down truck, and process stoppage can result in no stock. The manufacturer's production line stops, albeit for a short time.

Many factors have to be balanced to keep a supply chain running smoothly, managing risks, keeping utilization rates high, and reducing inventories, especially when the corporate mandate is to keep the end-product manufacturer always satisfied.

The politics of 'always satisfy the final-customer' and 'use the best SCRM management tools' can conflict. What does an operations manager do if there is a corporate mandate: 'never stock out of a product.' The end-product manufacturer may need the product to mitigate health, safety, or environmental risks. Or, the company needs the product for final production. Or, it is a high margin product. The company does not want to lose a single sale or face litigation. Ouch. Well, there has to be some buffer product inventory or approved alternate suppliers somewhere in the supply chain who can supply product on-demand. But, these emergency products come at a premium price.

SCRM FUTURE CAREERS

Two roads diverged in a wood, and I –
I took the one less traveled by,
And that has made all the difference.
Robert Frost

Executive management is aware that SCRM can generate revenue and lower costs. SCRM is growing exponentially since we live in a VUCA world moving at VUCA speed.

TECHNOLOGY ABILITIES GAP

Suppliers want to move up the SCRM process capability and maturity curve. There is only one problem. SCRM stakeholders such as operations managers, planners, manufacturing professionals, quality engineers and others cannot keep up with technology and the velocity of change. At its simplest, too many supply management people are still product and trans-actional focused, not process and risk oriented (RBPS).

SCRM AI and machine learning are also coming. They will facilitate and automate RBPS and RBDM. However, certain SCRM tasks will never be automated because they rely on individual judgment, discretion, and deep knowledge. For example, supply development, collaboration strategies, 'make or buy' decisions, risk-controls, corrective/preventive actions, and continuous innovation cannot be deferred to a software program. Technology may facilitate decision making, but it will not replace people judgment (RBDM).

SCRM FUTURE

When a management tool, idea, or philosophy is over promoted, there is a good chance it may become a fad. End-product manufacturers try it and lose interest when they do not receive instant gratification.

End-product manufacturers are developing SCRM. Suppliers will have to architect, design, deploy, and assure some basic level of SCRM, either having minimal processes or having 'world-class' processes for end-product manufacturers in highly competitive sectors. As well, companies will have to improve their quality, cost, technology, delivery, risk, and service processes by moving up the capability and maturity curve.

So, "Is SCRM here to stay?" We think SCRM will flourish if supply core processes are aligned and integrated with internal competencies. The corporate desire is to develop world-class, SCRM practices but people capabilities are frankly not up to the challenge. For too many years, operations professionals, quality professionals, engineers, purchasing agents, buyers, planners, and others have learned and deployed tactical tools.

YOUR SCRM FUTURE

SCRM matters to everyone from the Board to the CEO. Without a reliable flow of supplied products and services, there are no products to deliver on time, on budget to satisfied final-customers. Some supply chain pundits now say that SCRM efficiencies have the greatest potential to impact the bottom-line.

In fact, strategic sourcing and supply chain management have attained such high-level corporate visibility, that more people are moving from finance, operations, engineering, and information technology to SCRM.

Supply chain management is a hot career because it involves many functions such as quality, design, inventory management, production, logistics, storage, transportation, purchasing, finance, operations, marketing, sales, and even social media.

If you are involved with supply chain issues, this is your time. Many buying, logistics, and materials people have been in lower level, operational and management positions. SCRM is now a competitive strategic issue. This is a great time to make a difference in your career and to your company.

INDEX

ENDNOTES

[1] 'Steps to Building Supply Chain Resilience', *Supply Chain*, October 12, 2016.

[2] '3 Supply Chain Risks That Will Get You Thinking Bi-Modal', *Kinaxis*, December 7, 2016.

[3] 'A Gathering Storm of Political Risk Around the World', *Seeking Alpha*, October, 6, 2016.

[4] 'The Great Rewrite', *Forbes*/KPMG, 2016.

[5] **Supply Management Strategies for Improved Performance**, Greg Hutchins, 2002. P. 15.

[6] Disruption: An Opportunity to Excel, Business World, May 8, 2017.

[7] "Emerging Drought: Are Cities Planning for the Crisis? – Brink at the Edge of Risk, September 15, 2016.

[8] 'How Easy Is It for Criminals to Find the Weakest Link in Your Digital Supply Chain', *The Loadstar*, April 10, 2016.

[9] 'Abbott Recalls 465,00 Pacemakers for Cybersecurity Patch', *States News Service*, September 1, 2016.

[10] 'IOT Security Risks Begin with Supply Chains', *GovTech Works*, July 12, 2017.

[11] '3D Printing Blows Up Supply Chain Risk Management', *SDC Executive*, August 17, 2016.

[12] 'KPMG Voice: The Great Rewrite,' *Forbes*, 2016.

[13] ' BMW's Production Problems Highlight importance of Closer Collaboration', *The Engineer*, May 31, 2017.

[14] 'Hyundai Stops Operations of 4 Plants in China Due to Supply Problems', *Korea News Gazette*, August 31, 2017.

[15] 'Samsung's Other Battery Supply Chain Problem', *EPS News*, September 23, 2016.

[16] 'What Does It Take to Shut Down a Supply Chain', *MH&L*, August 16, 2017.

[17] 'To Save Retail Let It Die', *Retail Prophet*, September 17 2017.

[18] 'To Save Retail Let It Die', *Retail Prophet*, September 17 2017.

[19] What Risks Would a Trade War Bring?', *CFO Magazine*, March 6, 2018.

[20] What Risks Would a Trade War Bring?', *CFO Magazine*, March 6, 2018.

[21] Tradewinds, *BNP media*, October 2008.

[22] 'Supply Chain Using a Rating System,' *International Journal of Research in IT, Management, and Engineering*, August, 2016.

[23] IIA, 'Standards for the Professional Practice of Internal Auditing', Glossary, November 14, 2000.

[24] ISO 31000: Risk Management Principles and Guidelines, 2009.

[25] ISO 31000: Risk Management Principles and Guidelines, 2009.

[26] 'Risk Management', Business Dictionary, online, 2016.

[27] *American Society of Transportation and Logistics*, 2010, p. 69, 70.

[28] **APICS Basics of Supply Chain Management**, *Chron*, Small Business website, SmallBusiness.chron.com, 2015.

[29] 'Drill Down,' *VAR Business*, October 2008, September 9, 1997, p. 139.

[30] 'The Ripple Effect: How Manufacturing and Retail Executives View the Growing Challenge of Supply Chain Risk', *Deloitte*, 2013.

[31] **APICS Dictionary**, 13th edition, definition, APICS.org website, 2016.

[32] 'Unleashing Supply Chain Potential', *International Tax Review*, March, 1999, p. 28.

[33] 'Dealing With Supply Chain Risks: Linking Risk Management Practices and Strategies to Performance', *International Journal of Physical Distribution & Logistics Management*, p. 42, 2012.

[34] 'Grounded Definition of Supply Risk,' Zsidisin, Husdal.com website, 2016.

[35] Supply Chain Risk Management Practices for Federal Information Systems and Organizations, NIST 800 - 161, 2015.

[36] Best Practices and Supply Chain Risk Management for the U.S. Government, 2016.

[37] 'Supply Chain Risk Management: Compilation of Best Practices', August 2011, *SCRLC*.

[38] **Supply Chain Risk Management: Compilation of Best Practices**, Mitre Corporation, 2013.

[39] 'Managing Risk To Avoid Supply Chain Breakdown', *MIT Sloan*, Fall, 2004.

[40] **Paradigms: The Business of Discovering the Future**, Barker, Joel, Harper Business, 1992, pp. 32.

[41] Originated in Hutchins's **Risk Based Thinking**.

[42] 'Supply Chain Risk Management', Dun & Bradstreet, website, 2018.

[43] 'What Does It Take to Shut Down a Supply Chain', *MH&L*, August 16, 2017.

[44] 'The Seven Principles of Supply Chain Management', *Supply Chain Management Review*, Spring 1997.

[45] 'The Future of Work: Career Evolution', *The Economist*, January 29, 2000, p. 92.

[46] **Corporate Center**, Handy, Charles, *Executive Excellence*, December, 1998.

[47] **Beyond Certainty: The Changing Worlds of Organizations**, Handy, Charles, *Harvard Business School Press*, 1996, pp. 23-33.

[48] **Working It**, Greg Hutchins, QPE, 2001.

[49] 'Ghost Cars. Ghost Brands', *Forbes*, April, 30, 2001, p. 106.

[50] 'Sara Lee's Plan to Contract Out Work Underscores Trend Among US Firms', *Wall Street Journal*, March 17, 1997, p. A3.

[51] 'British Airways Moves Closer to Being Virtual Airline', *Wall Street Journal*, June 25, 1998, p. B1.

[52] 'Defining Challenge: Corporate American Confronts the Meaning of a Core Business', *Wall Street Journal*, November 11, 1999, p. A1.

[53] Derived from Greg Hutchins, **ISO 9000**, 2nd Edition, NY: John Wiley, 1996.

[54] **Strategy Pure and Simple**, Robert, Michel, New York: McGraw Hill, 1993.

[55] 'Making Companies Efficient', *Economist*, December 21, 1996, pp. 87-99.

[56] 'Outsourced But Not Outclassed,' *Computerworld Premier 100*, November 16, 1998, p. 36-37.

[57] 'Strategic Principle For Competing in the Digital Age', *McKinsey Quarterly*, October 18, 2016.

[58] 'Concurrent Development and Strategic Sourcing: Do the Rules Change in Breakthrough Innovation?', *Journal of High-technology Management Research*, Spring, 2000.

[59] 'Gucci's Surge and The Power of Responsive Supply Chain', *Business of Fashion,* November 1, 2017.

[60] 'Strategic Principles For Competing in the Digital Age', *McKinsey Quarterly*, October 18, 2016.

[61] 'Kellogg's You Supply Chain Model Comes at a High Cost', *Supply Chain Dive*, June 22, 2017

[62] 'GE Opening Micro Factory in Chicago to Build Industrial Prototypes', *Chicago Tribune*, October 19, 2016.

[63] 'Supply Chain Risk Management Approaches Under Different Conditions of Risk, *Inside Washington*, November 17 2014,

[64] 'Manager's Guide to Supply Chain Management', Stuart, Ian and McCutcheon, David, *Business Horizons*, March/April, 2000, p. 35.

[65] 'What Are Quality Standards for Experiences', Airbnb Website, 2018.

[66] 'More than a Feeling: Ten Design Practices to Deliver Business Value', *McKinsey*, December, 2017.

[67] The four attributes are derived from the definition in B. Enis, **Marketing Principles**, Glenview, IL: Scott, Foresman, 1980, p. 5.

[68] 'More than a Feeling: Ten Design Practices to Deliver Business Value', *McKinsey*, December, 2017.

[69] **Managing Quality**, D. Garvin, NY: Free Press, 1988, pp. 49-69.

[70] 'Tour the Software Chamber of Horrors', *Technology Review*, June, 1994, p. 6.

[71] 'Beyond the Supply Chain', *Industry Week*, November 2, 1998, p. 6.

[72] 'Manager's Guide to Supply Chain Management', Stuart, Ian and McCutcheon, David, *Business Horizons*, March/April, 2000, p. 35.

[73] 'One Thing Is Certain: 2017 Will Be Year of Uncertainty for CEO's', *Wall Street Journal*, December 27, 2016.

[74] ISO 9001: Quality Management System Requirements, ISO 9001:2015, p. ix.

[75] 'Cyber Threats and Risk Management When We're All Connected', *Wall Street Journal*, June 18, 2016.

[76] 'UPS Launches United Problem Solvers Campaign', *AdAge*, March 9, 2015

[77] 'Risk Appetite: Strategic Balancing Act,' EY Website, 2016.

[78] American National Standard Institute, Quality Management Systems – Fundamentals and Vocabulary, ANSI/ASQ Q9000 – 2000, p. 19.

[79] American National Standard Institute, Quality Management Systems – Fundamentals and Vocabulary, ANSI/ASQ Q9000 – 2000, p. 19.

[80] The PPRR Risk Management Model, Business Queensland, Australia, website, 2018.

[81] 'The New Economic Realities in Business,' *Management Review*, January, 1997, p. 13.

[82] Donald Rumsfeld, February 12, 2002 Department of Defense press conference.

[83] One Thing is Certain: 2017 Will be Year of Uncertainty for CEO's' *Wall Street Journal*, December 27, 2016.

[84] 'Supply Chain Fundamentals', *Modern Materials Handling*, February, 2001, p. 2.

[85] COSO - Internal Control - Integrated Framework, 2002.

[86] Adapted from **ISO 31000**, Greg Hutchins, 2016.

[87] 'Got a Good Idea for a New Product,' *Wall Street Journal*, May 1, 1997, p. 1.

[88] **Market Driven Strategy**, Day, George, NY: The Free Press, 1990, pp. viii-x.

[89] **Visionary Leadership**, San Francisco, Jossey Bass, Burt Nanus, 1992, p.8.

[90] 'Integrated Supply Chains: How to Make Them Work!', *Sourcing*, May 22, 1997, pp. 32-37.

[91] 'Leaders: Keeping the Customer Satisfied', *The Economist Magazine*, July 14, 2001, p. 9.

[92] **Management Fundamentals**, G. Dessler, Reston, VA: Reston Publishing Co., 1985, pp. 30-31.

[93] 'Top 10 Supply Risk Management Mistakes', KPMG, *Forbes Insights*, 2017.

94 IIA.

[95] 'Managing Risk in the Global Supply Chain', J. Paul Dittmann, June 4, 2014.

[96] 'Benefits and Risks of Outsourcing', *Lexology*, April 6, 2010.

[97] 'Five Top of Mind Matters for Supply Chain Risk Management', *Inside Counsel Magazine*, October 30 2014.

[98] 'Protecting Critical Assets from Cybertheft,' *Wall Street Journal*, October 17, 2016.

[99] 'Identifying Supply Chain Risks', Queensland Government website, 2016.

[100] 'How Supply Chain Management Affects a Company's Risk Management Strategy', *Logistics and Materials Handling*, September 29, 2016.

[101] 'Risk Appetite: Strategic Balancing Act', EY Website, 2016.

[102] 'Risk Appetite: Strategic Balancing Act', EY Website, 2016

[103] '3M Sees Opportunity in Supply Chain Risk,' *Risk & Compliance Journal*, June 15, 2017.

[104] 'Contain Global Supply Chain Risk', *Industry Week*, September 5, 2016.

[105] COSO, Key Concepts: COSO Definition of Internal Control, COSO Home Page, 2017.

[106] International Organization for Standardization, ISO/IEC 17000:2004 (ISO 170000, 2004).

[107] 'Supply Chain Managers: Are You the Biggest Risk In Your Supply Chain?', *Industry Week*, August 30, 2016.

[108] International Organization for Standardization, ISO/TR 31004/2013, Risk Management – Guidelines For the Implementation of ISO 31000.

[109] 'The Fine Tuned Organization', *Quality Progress*, February, 1992, pp. 47-48.

[110] **Corporate Center**, Charles Handy, *Executive Excellence*, December, 1998.

[111] 'The Future of Work: Career Evolution', *The Economist*, January 29, 2000, p. 90.

[112] **The Rebirth of the Corporation**, D. Quinn Mills, John Wiley & Sons, 1991.

[113] 'Management Discovers the Human Side of Automation', *Business Week*, September 29, 1986, p. 71.

[114] 'Re-inventing Yourself', *Industry Week*, November 21, 1994, pp. 20-24.

[115] Kobe Steel Quality Scandal Driven By Pursuit of Profits and Demanding Corporate Culture, *The Telegraph*, November 10, 2017.

[116] 'Guidance on Managing Outsourcing Risk', The Federal Reserve System, 2013.

[117] 'Daimler Warns of Supply Chain Risk From Switch to Electric Cars', Automotive News, February 14, 2018

[118] 'The Personal Commitment to Quality', *HR Magazine*, March, 1993, p: 100-101.

[119] 'Supply Chain Risk Management Approaches', *Inside Washington*, November 17 2014,

[120] Adapted from Total Quality Management - A Guide for Implementation, Superintendent of Documents, DOD 5000.51G, 1989, p. 15.

[121] Adapted from Total Quality Management - A Guide for Implementation, Superintendent of Documents, DOD 5000.51G, 1989, p. 15.

[122] 'Five Steps to Building Supply Chain Resilience', *Supply Chain*, October 12, 2016.

[123] 'The Next Generation Operating Model for the Digital World,' *McKinsey*, January 28, 2017.

[124] Supply Chain Risk Management Practices for Federal Information Systems and Organizations, NIST 800 - 161, 2015.

[125] '5 Critical Supply Risk Management Principles for Your Sourcing Process', *Spend Matters*, May 9, 2016.

[126] 'Risk: Not Resilience: Why Companies Avoid Examining Disruption', *Supply Chain Dive*, December 18, 2016.

[127] 'Airbus Warns It May Need to Stockpile Parts Against Brexit', *BBC News*, March 5, 2018.

[128] 'The Rewards of Best in Class Supply Risk Management', *Ardent Partners*, 2014.

[129] 'Don't Play It Safe When It Comes To Supply Chain Risk Management', *Accent Strategy*, August 20, 2016.

[130] 'One Thing is Certain: 2017 Will be Year of Uncertainty for CEO's', *Wall Street Journal*, December 27, 2016.

[131] ABC's of Conformity Assessment, NIST Special Publication, 2000-01, 2017. p. 6.

[132] 'Measuring Supplier Performance', *Business Journal*, June 4, 1999, p. 2.

[133] 'Another Jack Welch Isn't Good Enough', *Wall Street Journal*, November 22, 1999, p. A22.

[134] 'Carbide CEO to Forfeit Pay if Goal is Missed," *Wall Street Journal*, September 25, 1997, p. A3.

[135] 'Work Week', *Wall Street Journal*, April 6, 1999, p. A1.

[136] 'More Healthcare Suppliers Engage in 'Risky Business': 8 Things to Know About Risk Based Contracting', *Becker's Hospital Review*, March 8, 2018.

[137] 'How Chrysler Will Cut Costs', Sourcing Boston, February 8, 2001, pp. 30-32.

[138] **Built to Last**, Collins, James and Porras, Jerry, NY: Random House, 1994.

[139] Kobe Steel – Quality Trap, *Nikkei Asian Review*, October 27, 2017.

[140] 'Samsung Recall Puts Supply Chain Oversight in Spotlight', *Nasdaq*, October 11, 2016.

[141] 'Risk Mitigation Should be Part of a Healthy Supply Chain', *SDC Exec*, June 27, 2016.

[142] American National Standard, Quality Management Systems – Fundamentals and Vocabulary, ANSI/ASQ Q9000 – 2000, p. 6.

[143] **ISO 9000**, Hutchins, Greg, Vermont: Essex Junction, *OMNEO*, 1993, pp. 145-177.

[144] **Sourcing Strategies for Total Quality**, Hutchins, Greg, Homewood, IL: Business One Irwin, 1992, p. 15.

[145] 'For Automotive Purchasers ... The System is the Thing', *Sourcing*, Boston, February 11, 1999, p. 60.

[146] 'It's Not the Product That's Different, It's the Process', *NY Times*, December 15, 1999, p. C14.

[147] ASQ Survey, Despite Adversity in the Past, Manufacturers are Confident in *Supply Chain*, December 2016.

[148] 'A Sign of Altered Supply Chains', *Wall Street Journal*, August 29, 2016.

[149] 'DOD Needs Complete Information on Single-sources of Supply to Proactively Manage the Risks', Government Accountability Office, September, 2017.

[150] 'The Manager's Guide to Supply Chain Management', *Business Horizons*, March/April, 2000, p. 35.

[151] 'Why Don't Companies Look Closer at Disruptions,' *Strategic Sorcerer*, December 18, 2016.

[152] 'More Healthcare Suppliers Engage in 'Risky Business': 8 Things to Know About Risk Based Contracting', *Becker's Hospital Review*, March 8, 2018.

[153] 'Minimizing Supply Chain Risk', *Inbound Logistics*, April 2013.

[154] 'More Healthcare Suppliers Engage in 'Risky Business': 8 Things to Know About Risk Based Contracting', *Becker's Hospital Review*, March 8, 2018.

[155] 'ISO 9001:2015, Qualit Management Systems – Requirements, 2015.

[156] ANSI/ASQC, "Quality Systems - Model for Quality Assurance in Design, Development, Production, Installation, and Servicing - Q9001-1994, p. 1.

[157] 'Another Link in the Change', *Director*, March, 2000, p. 80.

[158] 'Protecting Your Supply Chain in the Face of Potential Auto Industry Challenges in 2018', *Monday Business Briefing*, March 7, 2018.

[159] FDA CFR Title 21, 820.1, Quality System Regulation.

[160] American National Standard Institute, Quality Management Systems – Fundamentals and Vocabulary, ANSI/ASQ Q9000 – 2000, p. 19.

[161] 'The Manager's Guide to Supply Chain Management,' *Business Horizons*, March/April, 2000, p. 35.

[162] This chapter was excepted from Hutchins, Greg, **Sourcing Strategies for Total Quality**, NY: Dow Jones Irwin, 1992.

[163] 'Concurrent Development And Strategic Outsourcing: Do The Rules Change In Breakthrough Innovation?', *Journal of High-technology Management Research*, Spring, 2000.

[164] 'Farm-to-Fork Food Safety Through an Effective Supplier Risk Management Strengthening Program', *ALM Media Properties*, June, 6, 2015.

[165] 'World Food Supplies At Risk As Threatens International Trade, Warn Experts', *Independent*, June 26, 2017.

[166] 'Change Threatens International Trade, Warn Experts', *Independent*, June 26, 2017.

[167] 'Change Threatens International Trade, Warn Experts', *Independent*, June 26, 2017.

[168] 'Climate Change Could Pose Significant Risk to National Security, Reports Say', *ABC News*, September 14, 2016.

[169] 'Climate Change Could Pose Significant Risk to National Security, Reports Say,' *ABC News*, September 14, 2016.

[170] 'Climate Change May Cut the World's Coffee Supply by Half', *Nature World News*, September 23, 2016.

[171] 'Farm-to-Fork Food Safety Through an Effective Supplier Risk Management Strengthening Program', *ALM Media Properties*, June, 6, 2015. David Acheson

[172] 'Curtailing Supply Chain Risks of Seafood Fraud', *Wall Street Journal*, September 12, 2016.

[173] What Does It Take to Shut Down a Supply Chain, MH&L, August 16, 2017.

[174] **Rapid Development**, McConnell, Steve, Microsoft Press, 1996, p. xiii.

[175] 'New Rules of Business', *Fast Company*, The Greatest Hits, Volume 1, p. i.

[176] 'Fear of the Unknown', *The Economist*, December 4, 1999, p. 61.

[177] 'Fear of the Unknown', *The Economist*, December 4, 1999, p. 61.

[178] 'Built To Last', *Sales & Marketing Management*, August, 1997, pp.78-83.

[179] 'GM is Seeking to Speed Up Development', *Wall Street Journal*, August 9, 1996, p. A3.

[180] 'Customer's Don't Want Choice', *Wall Street Journal*, April 18, 1998, p. A12.

[181] 'Bold and Costly Blunders', *Across the Board*, June, 1998, pp. 43-48.

[182] 'New Toothbrush Is Big-Ticket Item', *Wall Street Journal*, October 27, 1998, p. B1.

[183] 'Buyers Look to Distributors for Supply Chain Services', *Electronic Business*, February, 2000, p. 51.

[184] 'Behind the Wheel', *Wall Street Journal*, November 18, 1996, p. R14.

[185] 'Toyota Develops a Way to Make a Care Within Five Days of a Custom Order', *Wall Street Journal*, August 6, 1999, p. A4.

[186] 'Nissan to Offer Advance Alert System for Predicting Vehicle Maintenance and Identifying Defects', *IHS Global Insight*, September 6, 2017.

[187] 'Mitigating Medical Device Risks One of the Biggest Challenges of IT Pros,' Deloitte survey, *SCM Magazine*, August 18, 2017.

[188] Mitigating Medical Device Risks One of the Biggest Challenges of IT Pros, Deloitte survey, *SCM Magazine*, August 18, 2017.

[189] 'Is the Internet of Things Becoming an Internet of Risk', *IoT Agenda*, December 23, 2016.

[190] 'More than a Feeling: Ten Design Practices to Deliver Business Value,', *McKinsey*, December, 2017.

[191] 'Talking Toasters: Companies Gear Up For Internet Boom In Things That Think', *Wall Street Journal*, August 27, 1998, p. A1.

[192] 'Leaders: Keeping the Customer Satisfied', *The Economist Magazine*, July 14, 2001, p. 9.

[193] 'How Harley-Davidson and other Companies Deliver Individualized Results,' Digi-talist Magazine, SAP, July 2016.

[194] 'Bumper Crop: Competition Rises, Car Prices Drop: A New Golden Age?', *Wall Street Journal*, January 9, 1998, p. A1.

[195] 'Motorola Divides Semiconductor Sector in Five Groups to Hasten New Products', *Wall Street Journal*, May 28, 1997, p. B16.

[196] 'Gucci's Surge and The Power of Responsive Supply Chain', *Business of Fashion,* November 1, 2017.

[197] 'Three Elements of a Truly Effective Supply Chain Risk Management Strategy', *Llamasoft*, March 29, 2017.

[198] 'Buyers Look to Distributors for Supply Chain Services', *Purchasing Magazine*, February 10, 2000, p. 50.

[199] 'Think Demand, Not Supply', *Electronic Buyers' News*, October 10, 1997, p. 60.

[200] **Supply Chain Management: A Growing Field**, Winnipeg Free Press, June 6, 2017.

[201] 'Gucci's Surge and The Power of Responsive Supply Chain', *Business of Fashion,* November 1, 2017.

[202] 'Walmart Insistent on Efficiency: Charging Suppliers Late Early, or incomplete Orders', *The Toronto Star*, July 14, 2017.

[203] 'How Harley-Davidson and other Companies Deliver Individualized Results,' *Digitalist Magazine*, SAP, July 2016.

[204] 'Leaders: Keeping the Customer Satisfied', *The Economist Magazine*, July 14, 2001, p. 63.

[205] 'Just-in-time Systems', *Ward's Auto World*, May, 1999, p 67.

[206] 'Japanese Automakers, U.S. Suppliers and Supply chain', *Sloan Management Review*, Fall 2000, p. 81.

[207] 'Lean Manufacturing Process Needed for Survival', *Fort Worth Business Press*, 09/01/2000, p. 2A.

[208] 'Kaizen Costing for Lean Manufacturing', Lockwood, Diane L. and Modarress, Diane L, 2005, p. 57-61.

[209] 'Japanese Automakers, U.S. Suppliers and Supply chain', *Sloan Management Review*, Fall 2000, p. 81.

[210] 'Six Sigma', *Computerworld*, March 5, 2001, p. 38.

[211] **Benchmarking: A Tool for Continuous Improvement**; C. J. McNair and Kathleen H. J. Leibfried, 1993.

[212] **Benchmarking: The Search for Industry Best Practices that Lead to Superior Performance**, Camp, Robert, Milwaukee, WI: Quality Press, 1989.

[213] 'Drill Down', *VARBusiness*, September 9, 1997, p. 139.

[214] **APICS Dictionary 9th Edition**, APICS, Cox, James, 1998.

[215] 'Walmart Standard for Suppliers', Walmart web site, 2016.

[216] 'Walmart Standard for Suppliers', Walmart web site, 2016.

[217] 'France Drafts Law on Supply Chain Responsibility', *Just Style*, December 18, 2016.

[218] 'Major Food Companies To Tackle Water Risks in Their Global Supply Chains', *3BL Media*, October 25, 2016.

[219] 'Diageo, PepsiCo Make the Connection on Food Water Risk', *GreenBiz*, October 25, 2016.

[220] 'No Company Wants Slavery In Its Supply Chain, So Why Are So Many Sitting On Their Hands', *Thomson Reuters*, October 3, 2016.

[221] Security Management Systems for the Supply Chain, ISO/DIS 28001, 2007.

[222] "BMW to Recall 11,700 Cars After Installing Wrong Engine Software', *PM News*, February 26, 2018.

[223] 'Managing Information Security Risk, NIST 800 – 39, March 2011.

[224] Supply Chain Risk Management Practices for Federal Information Systems and Organizations, NIST 800-161, 2015.

[225] 'Solectron Becomes a Force in 'Stealth Manufacturing', *Wall Street Journal*, August 18, 1998, p. B4.

[226] The Ripple Effect, How Manufacturing And Retail Executives View The Growing Challenge Of Supply Chain Risk', 2012, *Deloitte Survey*.

[227] 'Change Management the Key to Supply Chain Management Success', *Automatic I.D. News*, April, 99, p. 40.

90342555R00239

Made in the USA
San Bernardino, CA
18 October 2018